Nuts and Bolts of Chemical Education Research

ACS SYMPOSIUM SERIES **976**

Nuts and Bolts of Chemical Education Research

Diane M. Bunce, Editor
The Catholic University of America

Renée S. Cole, Editor
University of Central Missouri

Sponsored by the
ACS Division of Chemical Education, Inc.

American Chemical Society, Washington, DC

Library of Congress Cataloging-in-Publication Data

Nuts and bolts of chemical education research / Diane M. Bunce, editor, Renèe S. Cole, editor ; sponsored by the ACS Division of Chemical Education, Inc.

    p. cm.—(ACS symposium series ; 976)

    Includes bibliographical references and index.

    ISBN 978–0–8412–6951–4 (alk. paper)

    1. Chemistry—Study and teaching—Research.

    I. Bunce, Diane M. II. Cole, Renèe S.

QD40.N88   2007
540.71—dc22

                                      2007060791

The paper used in this publication meets the minimum requirements of American National Standard for Information Sciences—Permanence of Paper for Printed Library Materials, ANSI Z39.48–1984.

ISBN 978-0-8412-6958-3 (paper)

Copyright © 2008 American Chemical Society

Distributed by Oxford University Press

All Rights Reserved. Reprographic copying beyond that permitted by Sections 107 or 108 of the U.S. Copyright Act is allowed for internal use only, provided that a per-chapter fee of $36.50 plus $0.75 per page is paid to the Copyright Clearance Center, Inc., 222 Rosewood Drive, Danvers, MA 01923, USA. Republication or reproduction for sale of pages in this book is permitted only under license from ACS. Direct these and other permission requests to ACS Copyright Office, Publications Division, 1155 16th Street, N.W., Washington, DC 20036.

The citation of trade names and/or names of manufacturers in this publication is not to be construed as an endorsement or as approval by ACS of the commercial products or services referenced herein; nor should the mere reference herein to any drawing, specification, chemical process, or other data be regarded as a license or as a conveyance of any right or permission to the holder, reader, or any other person or corporation, to manufacture, reproduce, use, or sell any patented invention or copyrighted work that may in any way be related thereto. Registered names, trademarks, etc., used in this publication, even without specific indication thereof, are not to be considered unprotected by law.

PRINTED IN THE UNITED STATES OF AMERICA

Foreword

The ACS Symposium Series was first published in 1974 to provide a mechanism for publishing symposia quickly in book form. The purpose of the series is to publish timely, comprehensive books developed from ACS sponsored symposia based on current scientific research. Occasionally, books are developed from symposia sponsored by other organizations when the topic is of keen interest to the chemistry audience.

Before agreeing to publish a book, the proposed table of contents is reviewed for appropriate and comprehensive coverage and for interest to the audience. Some papers may be excluded to better focus the book; others may be added to provide comprehensiveness. When appropriate, overview or introductory chapters are added. Drafts of chapters are peer-reviewed prior to final acceptance or rejection, and manuscripts are prepared in camera-ready format.

As a rule, only original research papers and original review papers are included in the volumes. Verbatim reproductions of previously published papers are not accepted.

ACS Books Department

Reviewers

Dr. Michael P. Doyle
University of Maryland

Alexander Grushow
Rider University

Gautam Bhattacharyya
Clemson University

William R. Robinson
Purdue University

Mary B. Nakhleh
Purdue University

Susan C. Nurrenbern
Purdue University

Catherine Milne
New York University

Donald Wink
University of Illinois

Xiufeng Liu
State University of New York at Buffalo

Thomas J. Greenbowe
Iowa State University

Thomas M. Holme
University of Wisconsin

Elizabeth Dorland
Washington University in St. Louise

Edward Grant
University of British Columbia

Contents

1. **Using This Book to Find Answers to Chemical Education Questions** 1
   Diane M. Bunce and Renée S. Cole

2. **Questions to Chemical Educators from the Chemistry Community** 11
   Richard N. Zare

3. **Funding Chemical Education Research** 19
   Loretta L. Jones, Maureen A. Scharberg, and Jessica R. VandenPlas

4. **Constructing Good and Researchable Questions** 35
   Diane M. Bunce

5. **Importance of a Theoretical Framework for Research** 47
   Michael R. Abraham

6. **The Particulate Nature of Matter: An Example of How Theory-Based Research Can Impact the Field** 67
   Vickie M. Williamson

7. **Qualitative Research Designs in Chemistry Education Research** 79
   Stacey Lowery Bretz

8. **Using Inferential Statistics to Answer Quantitative Chemical Education Research Questions** 101
   Michael J. Sanger

9. **Mixed Methods Designs in Chemical Education Research** 135
   March Hamby Towns

10. **Designing Tests and Surveys for Chemical Education Research** ... 149
    Kathryn Scantlebury and William J. Boone

11. Drawing Meaningful Conclusions from Education Experiments 171
    Melanie M. Cooper

12. Assessment of Student Learning 183
    Christopher E. Bauer, Renée S. Cole, and Mark F. Walter

13. Collaborative Projects: Being the Chemical Education Resource 203
    Barbara A. Sawrey

14. Buiding a Fruitful Relationship between the Chemistry and
    Chemical Education Communities within a Department
    of Chemistry 215
    Gabriela C. Weaver

Indexes

Author Index 229

Subject Index 231

Chapter 1

Using This Book to Find Answers to Chemical Education Research Questions

Diane M. Bunce[1] and Renée S. Cole[2]

[1]Chemistry Department, The Catholic University of America, Washington, DC 20064
[2]Department of Chemistry and Physics, University of Central Missouri, Warrensburg, MO 64093

> The perceptions and expectations of chemical education research by chemical education researchers, chemical researchers, chemistry teachers, and funding agencies are often very different from each other. This chapter provides an overview of the field for these different groups that includes a discussion of what constitutes quality chemical education research. A roadmap for this book is provided to help the reader find answers to specific questions including how to generate ideas to investigate, develop appropriate methodologies, assess student learning, find an agency or foundation to fund chemical education research, or collaborate with others in a joint project with a chemical education component.

© 2008 American Chemical Society

Introduction

Chemical education research (CER) is a relatively new field that combines the theories, experimental designs, and tools of several disciplines such as education, psychology, and sociology with the issues of teaching and learning chemistry. As such, chemical education research is often misunderstood within the field of chemistry. The expectations from outside CER of what chemical education research should be addressing, how it should operate, and what it should produce are often at odds with reality. Yet, chemical education research can significantly add to the knowledge of how we teach and the impact upon the learners of chemistry. As a field, chemical education research has the ability to answer questions that have plagued members of chemistry departments (Chapter 2, Zare) and funding agencies (Chapter 3, Jones, Scharberg, and VandenPlas) for generations. What is the best way to teach? What are the best topics to teach? Why don't students learn? How effective are alternate teaching approaches? How can we improve what we do?

There are many audiences with a vested interest in the answers to these questions including teachers, researchers, administrators, funding agencies, and students. Each of these stakeholders brings to these questions a different perspective and a somewhat different expectation of the answers. Chemical education research is the field of study that is able to address these questions with a systematic, logical, verifiable and convincing approach. Our goal in this book is to provide an overview of the components of chemical education research and to discuss the process of how questions in this field are addressed.

The audience for this book is diverse. It includes chemists who want to understand and evaluate the chemical education research literature for use in their own classes, grant writers who want to include quality chemical education research in their proposals, researchers trained as chemists who want to conduct chemical education research, new chemical education researchers who want to expand their understanding of the field, experienced chemical education researchers who want to review aspects of the research process, students who are studying chemical education research, and funding agencies and foundations who fund chemical education research through competitive grants. Authors provide references in each chapter to help the reader develop a more complete understanding of the topic.

Recurring Themes

Questions

There is a good deal of discussion in this book regarding questions that chemists believe chemical education research should investigate (Chapter 2,

Zare), that funding agencies want addressed (Chapter 3, Jones, Scharberg, and VandenPlas), that interest teachers (Chapter 12, Bauer, Cole, and Walters) and that theory frames (Chapter 5, Abraham). To investigate the question in a meaningful way, that question must be posited in a way that lends itself to a systematic investigation (Chapter 4, Bunce; Chapter 12, Bauer, Cole, and Walters). The process of moving from question of interest to researchable hypothesis is multi-step and requires a good deal of thoughtful consideration. If done correctly, many questions concerning the research design are answered by the question the researcher asks.

Chemical education research questions are developed by different stakeholders including funding agencies, collaborators from other fields (Chapter 13 Sawrey), and individual researchers, but the process of operationalizing that question contains some common elements (Chapter 4, Bunce). Among these elements are what constitutes evidence, what evidence will the audience of interest find convincing, and how will the answers to this question extend the theories that already exist about teaching and learning (Chapter 5, Abraham).

One aspect of developing questions that can be often overlooked at the onset of a project is how the results of the experiment will extend our knowledge of how teaching and learning take place and what makes them effective. If a question is asked in isolation, apart from a theory of learning and teaching, the end result is likely to be little more than an isolated fact that does not advance knowledge of teaching and learning. If repeated over time, this error generates several points of knowledge that are not interconnected. This, like the knowledge base of novice learners, can lead to a very limited view of the problem and its solution. If, on the other hand, the research question is framed within the context of a specific theory, the results of an individual experiment can be explained in the larger context of the theory. The end result is integrated with what we already know and the sum of this new and old knowledge can lead to more and better investigations of teaching and learning. The case study (Chapter 6, Williamson) of how the theory of the Particulate Nature of Matter shaped our understanding of how students learn or fail to learn the abstract molecular-based concepts of chemistry provides a clear example of this point. The end result is that teaching, curriculum, textbooks, and national examinations changed within a relatively short time based upon the research results from the investigations of the Particulate Nature of Matter theory.

Research Methodologies

The research methodology used to investigate fully operationalized questions must match the questions asked. It is not enough to say that a question was investigated using a treatment and control methodology if such a methodology masks the important underlying variables. Many educational experiments in the past have failed to yield convincing results, in part, because an inappropriate methodology was used to investigate the question. This situation is analogous to saying that the only acceptable methodology for characterizing a compound is with a UV spectrometer. Although a UV spectrophometer is the instrument of choice in some situations, it is not the best choice in all situations. The same is true of a treatment and control design in chemical education research.. In chemical education, as in chemistry, the methodology used to investigate a question must match the question asked.

One challenge regarding the diversity of methodologies in chemical education research is that they are not widely known within the chemistry community. Three chapters in this book (Chapter 7, Bretz; Chapter 8, Sanger; and Chapter 9, Towns) describe general approaches to research design in chemical education research, namely, qualitative, quantitative, and mixed methods designs. All three authors provide an informative overview of the respective methodology and compare and contrast it to the other two. Detailed bibliographies in each chapter guide the reader to more specific information.

Research Results

One of the most important ways we convey research is through the presentations and publications. A glance at the Chemical Education Division Program of any national American Chemical Society Meeting will show a large number of chemical education research presentations on the schedule. Although research journals that publish chemical education research are still small in number, the situation is changing. More journals and expanded research features in established journals that accept chemical education research are appearing.

As with all research, manuscripts must undergo peer review. Many authors who submit research to chemical education research features or journals are surprised when their manuscripts are critiqued for drawing conclusions unsupported by their data. Sometimes this is a result of not developing researchable questions or not selecting an appropriate methodology that controls for intervening variables. But in some cases, the critique centers on the construction of inappropriate conclusions based upon the data collected or analyzed. Cooper (Chapter 11) addresses this point and demonstrates how the use of a theoretical basis for an experiment can help interpret the significance of research results.

Human Subjects

A major difference between investigating molecules and people is that people have the right under law to agree or not agree to participate in the research. This right is protected by each institution's Institutional Review Board (IRB). The role of this board is to review the research *before it begins* and either determine that the research falls within standard practices and warrants an exemption or fully investigate the research plan to ascertain the discomfort, imposition or risk such research causes the participant. Many authors in this book have referred to the issues involved in gaining IRB approval for research (Chapter 3, Jones, Scharberg, and VandenPlas; Chapter 4, Bunce; Chapter 7, Bretz; Chapter 8, Sanger; Chapter 12, Bauer, Cole, and Walters; and Chapter 13, Sawrey).

Specific Issues

Assessment of Learning

Many readers of this book will be interested in chemical education research as a means of evaluating either their own teaching or some innovation used within their institution. Often this necessitates measuring changes in what students know or think. Chapter 12 (Bauer, Cole, and Walters) directly addresses the question of how to design an assessment of student learning, including an overview assessment questions, and a guide to appropriate tools to use for specific questions of student learning. Readers particularly interested in authentic or meaningful student assessment may wish to start with Chapter 12 and then in succession turn to the chapters on using theory to frame the research (Chapter 5, Abraham), writing questions (Chapter 4, Bunce), choosing methodologies (Chapter 7, Bretz; Chapter 8, Sanger, and Chapter 9, Towns), writing survey and test items (Chapter 10, Scantlebury and Boone) and using data appropriately to support conclusions (Chapter 11, Cooper) for more in-depth information.

Developing and Analyzing Tests and Surveys

Many investigators use tests and surveys as research tools. Research submitted for publication often includes entire research experiments that depend on a single test or survey. Yet the researchers fail to provide any evidence on how the survey or test was constructed and proven to be both valid and reliable. Basing the results of research on such an unproven instrument is akin to using a non-zeroed balance to weigh a valuable sample and then reporting the result to 5 significant figures. Tests and surveys that are used as research instruments must undergo the same "zeroing" (testing for validity and reliability) that any scientific instrument would. Chapter 10 (Scantlebury and Boone) guides the reader past the typical pitfalls of constructing tests and surveys. More importantly, these authors provide guidance on both traditional methods of analysis and the asch Method. The Rasch Method provides a way to reduce surveys to a minimum number of questions and determine the balance among hard and easy questions within the test or survey. It also can be used to combine data from administrations of slightly different forms of the same survey.

Writing a Grant Proposal

Knowing how to do chemical education research is one thing but without appropriate funding, even the most exquisite ideas and hypotheses can go unexamined. Chapter 3 (Jones, Scharberg and VandenPlas) provides a practical guide to finding an appropriate funding agency, and once identified, writing a suitable proposal. This chapter's extensive list of government agencies/programs and private foundations that fund chemical education research projects is a good place for any researcher to start. Advice on practical issues such as creating the budget will prove useful for the novice through expert investigator.

Collaboration in Grants with Educational Components

An interesting extension of funding chemical education research through grants is Chapter 13 (Sawrey) that examines how to develop and negotiate collaborative grant proposals that include an education component. This chapter explores the differing roles a chemical education researcher can play in such grants. Roles range from education specialist (curriculum designer) to content specialist (evaluator) . Sawrey points out the benefits and pitfalls of each role and discusses the type of collaborative grants that are possible.

Putting Chemical Education Research in Context

Chapters 2 (Zare) and 14 (Weaver) give two perspectives on chemical education research: the chemist's view of the questions that chemical education research could address and a chemical education researcher's view on how to grow and thrive within a chemistry department. These two authors essentially define the conversation that currently exists within chemistry departments on what chemical education research is capable of doing and how it can accomplish this within the scientific research community. To be successful, chemical education research must bridge both the culture and the practice of two distinct disciplines (chemistry and education) and do it within the language of scientific research. The authors in this book argue that the culture of scientific research must be expanded beyond the well-defined limits of traditional bench chemistry to empower the investigation of critical questions in the teaching and learning of chemistry. Just as chemistry is engaged more and more in using interdisciplinary approaches to answer questions, so too, chemical education research uses the theories, methodologies, and tools of other fields to investigate its critical questions. These modified theories, methodologies, and tools should be viewed in the same spirit of scientific inquiry as are the physics, biology, and electronic tools used in the pursuit of chemistry questions.

Road Map for Using this Book

Some readers may choose to read this book from cover to cover to in order to systematically develop or re-examine their understanding of chemical education research. Others may choose to use this book as a "how to" guide for a specific purpose. In order to facilitate both uses of the book, Table I points readers to specific chapters to address particular questions:

Table 1. Roadmap for Using the Book

Question	Chapter Title	Author
How do I find out if my students are learning?	Chapter 12: Assessment of Student Learning—Guidance for Instructors	Christopher F. Bauer, Renee S. Cole, Mark F. Walters
How is chemical education research different than just evaluating what goes on in the classroom?	Chapter 12: Assessment of Student Learning—Guidance for Instructors	Christopher F. Bauer, Renee S. Cole, Mark F. Walters
How do I come up with questions to investigate?	Chapter 3: Funding Chemical Education Research	Loretta L. Jones, Maureen A. Scharberg, Jessica R. VandenPlas
	Chapter 4: Constructing Good and Researchable Questions	Diane M. Bunce
	Chapter 5: Importance of a Theoretical Framework for Research	Michael R. Abraham
How do I get funding to do chemical education research?	Chapter 3: Funding Chemical Education Research	Loretta L. Jones, Maureen A. Scharberg, Jessica R. VandenPlas
	Chapter 13: Collaborative Projects—Being the Chemical Education Resource	Barbara A. Sawrey

How can I include an education component in a chemistry grant?	Chapter 13: Collaborative Projects—Being the Chemical Education Resource	Barbara A. Sawrey
How do I investigate the research questions once I have them?	Chapter 4: Constructing Good and Researchable Questions	Diane M. Bunce
	Chapter 5: Importance of a Theoretical Framework for Research	Michael R. Abraham
	Chapter 6: An Example of How Theory-Based Research Can Impact the Field	Vickie M. Williamson
	Chapter 7: Qualitative Research Designs in Chemistry Education Research	Stacey Lowery Bretz
	Chapter 8: Using Inferential Statistics to Answer Quantitative Chemical Education Research Questions	Michael J. Sanger

Continued on next page.

Table 1. *Continued.*

Question	Chapter Title	Author
How do I investigate the research questions once I have them?	Chapter 9: Mixed Methods Deigns in Chemical Education Research	Marcy Hamby Towns
	Chapter 10: Designing Tests and Surveys for Chemistry Education Research	Kathryn Scantlebury, William J. Boone
	Chapter 11: Drawing Meaningful Conclusions from Educational Experiments	Melanie M. Cooper
How can I revise my rsearch manuscript to make it acceptable for publication?	Chapter 5: Importance of a Theoretical Framework for Research	Michael R. Abraham
	Chapter 11: Drawing Meaningful Conclusions from Educational Experiments	Melanie M. Cooper
What is expected of me as a chemical educator by the chemistry community?	Chapter 2: Questions to the Chemical Educators from the Chemistry Community	Richard N. Zare
	Chapter 14: Building a Fruitful Relationship between the Chemistry and Chemical Education Communities within a Department of Chemistry	Gabriela C. Weaver

Chapter 2

Questions to Chemical Educators from the Chemistry Community

Richard N. Zare

HHMI Professor and Chair, Department of Chemistry, Stanford University, Stanford, CA 94305–5080

This chapter poses twenty questions whose answers are of vital importance in teaching chemistry to beginning students at the college or university level. These questions are addressed primarily to those who carry out research in chemical education, but the argument is made that the answers provided by this community of scholars will have little impact unless chemists and chemical education researchers can communicate clearly to one another and gain not only each other's respect but also the attention and respect of the wider chemistry community.

Chemical education research is similar to research in the chemical sciences. The investigator begins with a question, defines what needs to be better understood, designs experiments to collect data, analyzes the collected data using the most sophisticated tools available, and fully discloses the work in the form of refereed publications and conference proceedings. Reflection on the validity of the hypothesis compared with the observed findings creates new questions, requires modification of the original propositions, and so on. Yet, most chemists feel much more comfortable with research on chemical problems than research on chemical education problems. Why is that so?

I suspect it is because of the inherent complexity of chemical education problems. The evidence that chemists find compelling is usually quantitative rather than qualitative, and we tend to distrust experiments that cannot be exactly reproduced. It is easy to argue that presenting the same material in the same fashion in different classes will yield different results just because different students will be present, and this fact leads some chemists to scorn all efforts to investigate which teaching approaches are most effective. Chemists are drawn to the study of pure substances under conditions where the response of the chemical system results in a linear change with the experimenter's variations of the initial conditions. But research on teaching and learning is not like that.

Actually, the chemical world is not like that either. Chemists are increasingly aware that by avoiding complexity and heterogeneity they can miss important discoveries, such as the details of how living cells work. Thus, while chemists might be skeptical of chemical education research in the same way that they are skeptical about the social sciences, this research area is not only a valid one but one that holds huge potential for practical gains in preparing the next generation of chemists. Nothing is more fundamental to the future of the profession than attracting talented young women and men to the pursuit of the chemical sciences and providing them with an education adapted for solving problems at the cutting edge of our field.

Chemists and chemical education researchers have this goal in common, but it fails to unite their efforts. The findings of the two groups often are described in separate jargons and almost always published in separate journals. In a speech to the Northeast Section of the American Chemical Society, Dr. Robert L. Lichter, then Executive Director for the Camille and Henry Dreyfus Foundation, commented on this separation (*1*):

> There's a tendency to divide the chemical universe into two groups: the educators and the doers. Conferences and other gatherings on the topic [of education] tend to be directed to those called the former. I suggest that this is a highly limited perspective and does the profession and the practice, and certainly the students, a disservice.

I myself would divide the chemical universe into chemical researchers, chemical educators, and chemical education researchers. Only a few people belong to all three groups but many if not most people belong to two, while it cannot be denied that some people identify themselves as belonging to only one.

George Bernard Shaw wrote in "Maxims for Revolutionists," an appendix to his play *Man and Superman*, the infamous lines (*2*):

> He who can, does.
> He who cannot, teaches.

The corollary has been proposed (*3*):

> He who cannot teach, teaches teachers.

This painful put-down of teaching and research into understanding how students learn expresses a common attitude among chemistry faculty members in institutions of higher learning -- institutions where the integration of teaching and research remains more a mantra mumbled by administrators than a practice embraced by professors. The incentive system at research universities has historically rewarded scientists richly for making discoveries and publishing academic papers but poorly for nurturing students, some of whom will become the next leaders. Moreover, it is easy to construct metrics for measuring research productivity but much harder to do the same for teaching and mentoring. And what metrics are we to use for chemical education researchers? Clearly, this activity has many variables to handle, large questions to examine, and different tools to use in its experimental design, but it is commonly dismissed as a second class activity by many chemists at research universities.

Most faculty members originally became professors because they believe that teaching is a noble endeavor; teachers influence lives and shape futures. For many years chemists have exchanged ideas about effective teaching at meetings and in peer-refereed journals. Unfortunately, this activity is not regarded as a mainstream responsibility for all chemists who teach. I strongly endorse the sentiments expressed so eloquently by Coppola and Jacobs (*4*): "In general, the scholarship of teaching and learning shows great promise for enriching and supporting chemistry education because it seeks to make systematic, scholarly thinking about teaching and learning a part of every faculty member's life, rather than just those who have claimed its specialization."

What do we know about what makes a student choose chemistry as a career path? A consensus has emerged that undergraduates need early, engaging hands-on experiences in the laboratory and much more mentoring than most of them presently receive to maintain their interest and inspire them to take up careers in the sciences, if not chemistry. A means must be found to enliven a dry and dispiriting style of science instruction that leads as many as half of the country's aspiring scientists to quit the field before they leave college. Many, including me, feel that the nation's future is at risk without investing in better science and math education for the next generation (*5*). The time has come to ask chemical education researchers for their help in carrying out the heavy responsibilities of being a university chemistry professor. I want them to address questions whose answers will help chemistry professors apply sound and proven principles to their teaching.

What follows is a list of twenty questions that I would like to see addressed by chemical education researchers – but to which all chemical educators are invited to contribute. The list captures for me some (but not all) of the perplexing problems that chemistry instructors confront. These questions are put forward by someone who has taught beginning chemistry students at Stanford for nearly 30 years but has never received any formal training in chemical education and in no way considers himself a chemical education researcher:

1. What makes introductory chemistry courses so hard for students?

2. Why do some students steadily improve while others steadily decline in beginning chemistry courses?

3. How do we make chemistry courses about learning rather than about getting good grades?

4. What is the importance of lecture demonstrations?

5. What is the importance of the beginning laboratory experience?

6. How should we teach beginning chemistry students with widely different backgrounds?

7. How significant is a teacher's choice of a definite curricula?

8. How significant is teaching style?

9. What role should instructional technology play in teaching and learning?

10. How can beginning faculty members improve teaching skills?

11. What factors make undergraduates major in chemistry?

12. What is the right balance between teaching and research demands?

13. What aspects of teaching the chemical sciences are unique to chemistry?

14. What should we put in and what should we take out of the chemistry curriculum?

15. What are the advantages and disadvantages of team teaching to student learning?

16. What chemistry should we teach to non-science majors?

17. How important are group learning activities to student learning?

18. How important is it to develop the communication skills of students?

19. What should students know about using the chemical literature to become practicing chemists?

20. What are the successful strategies for solving chemical problems?

No claim is made that these questions are new ones. They have been and are being addressed by chemical education researchers, but the results are largely unknown to the greater chemistry community, because they are usually published in chemical education journals for an audience of chemical education researchers. This failure to communicate results to chemistry instructors adversely affects their ability to teach.

Let us examine one example of this communications failure. The *Journal of the American Chemical Society* (JACS) was founded in 1879 and is regarded to be the flagship journal of the American Chemical Society. JACS claims to be devoted to the publication of research papers in all fields of chemistry and publishes approximately 17,000 pages of new chemistry a year. You will find between its covers articles, communications to the Editor, book reviews, and computer software reviews. But, you will seldom if ever find anything in JACS about research in chemical education. The consequences are the institutionalization of a divide between chemists and chemical education researchers – a divide that prevents either group from seriously influencing the actions of the other. A litany of other such examples of peer-reviewed chemistry research journals being blind to chemical education research can be recited. Of course, the argument goes both ways. Some chemical education research articles are so full of jargon and so strongly focused on impressing other chemical education researchers that they are nearly impenetrable to chemists. The blame game is not interesting; doing something to promote communication between these two groups is truly valuable. I strongly advocate that these two communities must speak more to one another, or suffer the consequences of both being impoverished by this lack of information and opinion exchange.

I hasten to admit I am not sure of the answers to the twenty questions posed above, but I do have some thoughts. Many students begin my introductory chemistry course with a sense of dread, believing that the chemistry department is a gatekeeper that stands in the way of their achieving their aspirations, or often more correctly their parents' aspirations, that they become medical doctors. This sad situation is common in the United States, and it challenges many chemistry instructors. However, the same class contains students who will discover that the study of chemistry fires their imaginations and opens new possibilities that they never considered before. With all of this in mind, I have thought a great deal about the first of the twenty questions that I posed: why is an introductory chemistry course so hard for students? Most students experience a learning discontinuity between high school and college chemistry. The former frequently rewards memorization, recitation, and using algorithms to solve word problems, whereas the latter often demands reasoning from understood concepts. Many students work very hard in the same mode that was successful in high school chemistry only to discover that this approach is like hitting your head against the wall. All of us who teach introductory chemistry hope to find a way for students to come to this realization prior to receiving poor marks on exams. How do you do this?

My own approach has been to give many small exams called homework. Homework counts for very little of the final grade in the course, but these students have gotten to Stanford by always completing assignments. The homework assignments are important in communicating to them the type of skills that they must acquire to succeed in this course. I point out that no one learns how to play the piano by reading a book on how to play the piano. In the same way, working problems is what they need to insure that they have secured mastery of the course.

It is my experience that students who drop out of beginning chemistry do so because they fall behind, panic, and reach a state of mind where rational discourse and even intense intervention are futile. To prevent this state of collapse, I assign homework in each lecture that is due at the next lecture. I encourage students to work the problems on their own at first, then discuss them with classmates. Because I grade the course on an absolute basis, the students are not competing against each other for grades, and I encourage them to work together by assigning them to study groups. These groups are based on geographical proximity, taking into account what dorms the class members reside in. I find that self-selected study groups tend to leave out some class members. This conclusion is not original but builds on results obtained by many others.

More important than the opinions I presently hold on how the twenty questions might be answered is the fact that these opinions are subject to change. As I listen to others and reflect on the other chapters in this book, and as I try various approaches on my own students and observe the results, I sharpen and refine my own thinking on these questions. It is the quest that matters. An old Chinese proverb states (6): "Teachers open the door. You enter by yourself." But different people have different doors. What may be an open door for one student may be a wall for another. It is saddening to realize that no one correct set of answers probably exists to these questions. Conversely, multiple teaching approaches simultaneously available to the students of a class may open the most doors. Here is where chemical education research can show which of many different approaches works best for which student.

Certainly, it is important to try different teaching approaches. Experimentation – which always means risking failure – is at the root of almost every success. That adventurous spirit is required to succeed at developing new teaching methods, improving curricular content, and systematically testing which is best and why in which situation and with which student. Assuredly, teachers must honor the best of education's established practices, but they must not shy away from investigating new methods to reach students. And chemical education research can help us discover which methods work well. Instructional methodology must be perpetually evaluated and improved upon – or discarded as ineffective. Just as in research, what is needed in teaching is a spirit of playfulness combined with critical evaluation and assessment of the outcomes.

Let me return to George Bernard Shaw's quote that began this short chapter. I endorse Lee Schulman's sentiments when he wrote (7):

> "We reject Mr. Shaw and his calumny. With Aristotle we declare that the ultimate test of understanding rests on the ability to transform one's knowledge into teaching. *Those who can, do. Those who understand, teach.*"

to which I should add, and those who seek the connection between the two do chemical education research. To paraphrase how Schulman concluded his essay, I would write:

> Those who can, do.
> Those who cannot, do not.
> Those who can do, and who can teach and reflect on what makes teaching effective, do it all!

We still have so much to learn about teaching chemistry and the first step is asking good questions. But posing questions from the chemistry community to the chemical education research community is not enough. Unless both communities deepen their respect for each other and exchange more information and ideas between them, the answers provided by chemical education researchers are likely to fall on deaf ears.

References

1. Lichter, Robert L. Private Communication.

2. Shaw, G.B. "Maxims for Revolutionists," (1903). See http://www.bartleby.com/157/6.html
3. Peter, Laurence J. "Peter's Quotations: Ideas for our Time," William Morrow & Co. (1977).
4. Coppola, B. P. and Jacobs, D. "Is the Scholarship of Teaching and Learning New to Chemistry?" in, M. T. Huber and S. Morreales (Eds.), Disciplinary Styles in the Scholarship of Teaching and Learning. A Conversation. Washington DC: American Association of Higher Educaton and The Carnegie Foundation for the Advancement of Teaching, 2002; pp. 197-216.
5. "Rising Above the Gathering Storm: Energizing and Employing America for a Brighter Economic Future," NAS/NAE/IOM, Washington, DC, 2006.
6. http://www.quotationspage.com/quote/29226.html
7. Shulman, L. S. The Wisdom of Practice. Essays on Teaching, Learning, and Learning to Teach. Jossey-Bass: San Francisco, CA, 2004, p.212.

Chapter 3

Funding Chemical Education Research

Loretta L. Jones[1], Maureen A. Scharberg[2], and Jessica R. VandenPlas[3]

[1]Chemistry Program, University of Northern Colorado, Greeley, CO 80639
[2]Department of Chemistry, San Jose State University, San Jose, CA 95192
[3]Chemistry Department, The Catholic University of America, Washington, DC 20064

The key ingredients of successful proposal writing for chemical education research are addressed, including how to find an appropriate funding agency for a research project, responding to a call for proposals, and establishing a project budget.

Getting Started

The suggestions for research directions in Chapter 2 (*1*) present some of the important problems encountered in the learning of chemistry. Such suggestions can lead to fruitful research investigations; other ideas may come from one's own experiences in the classroom or from studying how learning of other disciplines has been enhanced. In any case, often a researcher may need funding to carry out an investigation, and thus must begin the process of locating and applying for a grant from a federal, state, or private agency. Typically, in chemical education, the types of grants sought fall into the following categories: educational research, equipment and materials for curriculum development and implementation, evaluation of curriculum innovation or materials, workshops, dissemination, travel and planning grants. Most likely, the proposed project will fall into one or more of these categories. Once one has identified the research he/she would like to pursue, they must determine how his/her research goals fit into the interests of a funding agency.

Locating a Funding Source

Before searching for a suitable granting agency, it is a good idea to write a concept outline of the project to help formulate the proposal. It is best to keep the outline to one page that can be sent to colleagues for their feedback and to program officers in potential funding agencies to assess their interest. Brief statements, bullets, and keywords are more useful than detailed descriptions at this point. An excellent sample concept outline can be found under Downloadable Forms on the Website of the Missouri State University Sponsored Programs Office (2).

To match a research project to a granting agency, a variety of resources can be consulted. A university research or contracts and grants office is a good starting point. They can help the researcher match a proposed research project with a funding agency. Colleagues experienced in chemical education research can also provide advice on potential granting agencies that are suitable for a given project. Additional resources for locating funding sources are available on the Internet. Table I lists the major federal agencies that have funded chemical education projects, while Table II lists some of the leading private agencies. State and regional agencies exist as well, but these are best identified by talking with local university research offices.

Each granting agency has its own criteria for funding, methods by which it reviews proposals, and types of projects that it will fund. In addition, the aims of funding agencies vary both over time and with different leadership. Private foundations must answer to changing boards of directors, and federal agencies to the various interests of the U.S. Congress. Finally, it is important to note that some granting agencies fund only individual projects while others fund

Table I. Federal agencies that have funded chemical education projects.

Federal Agency	*Website*
National Science Foundation (NSF)	http://www.nsf.gov
National Institutes of Health (NIH)	http://www.nih.gov/
Department of Education (DOE)	http://www.ed.gov/
Environmental Protection Agency (EPA)	http://www.epa.gov/
Office of Naval Research (ONR)	http://www.onr.navy.mil/

Table II. Private agencies that have funded chemical education projects.

Private Agency	Website
Research Corporation	http://www.rescorp.org/
Exxon Education Foundation	http://www.exxonmobil.com/Corporate/Citizenship/gcr_education_main.asp
Howard Hughes Medical Institute	http://www.hhmi.org/
The Coca-Cola Foundation	http://www2.coca-cola.com/citizenship/foundation.html
The Camille & Henry Dreyfus Foundation	http://www.dreyfus.org/
Bill & Melinda Gates Foundation	http://www.gatesfoundation.org/
The Kresge Foundation	http://www.kresge.org/
W. M. Keck Foundation	http://www.wmkeck.org/
The David & Lucile Packard Foundation	http://www.packard.org/
The William and Flora Hewlett Foundation	http://www.hewlett.org/
Carnegie Foundation for the Advancement of Teaching	http://www.carnegiefoundation.org/
Ford Foundation	http://www.fordfound.org/
The Annenberg Foundation	http://www.whannenberg.org/
Noyce Foundation	http://www.noycefdn.org/

collaborative projects as well. Collaborative projects might bring together faculty members within a department, college or division, or even expand to a variety of institutions nationally or internationally. They are often a good way to get started working with funded projects. Chapter 13 (*3*) describes the various roles that a chemical education researcher can play within a collaborative project and how to succeed in that role, while this chapter focuses on developing a proposal for which the chemical education researcher is the Principal Investigator.

Even within a granting agency, a researcher may find diverse criteria for funding and the types of projects that may be supported. A large federal agency, such as the National Science Foundation (NSF), the U.S. Department of Education (DOE), or the National Institutes of Health (NIH), consists of a collection of divisions or directorates, each of which may value different research questions. It is important to identify not only the agency that best fits the desired research project, but also the division within that agency that is most in line with the research goals. At NSF, for example, grants for research in education may fall into a number of different directorates within the Education and Human Resources Directorate (EHR). Comprehensive research grants are sought primarily by the Research and Evaluation on Education in Science and Engineering (REESE) Program, while grants for applied research are awarded through the Divisions of Undergraduate Education (DUE) and Graduate Education (DGE), as well as other divisions within EHR. Research questions that appeal to program officers in EHR as a whole might include the following general goals (*4*):

- What are the areas of importance to education research and practice?
- What rigorous methods can be developed for synthesizing findings and drawing conclusions?
- How can we advance discovery and innovation at the frontiers of science, technology, engineering, and mathematics (STEM) learning?

The National Science Foundation states that proposals to conduct research in science education "are expected to be based deeply in the STEM disciplines and be theoretically and methodologically strong with the potential of contributing to theory, methodology, and practice (*4*)."

Some research questions that have been funded by the REESE program within EHR are found in the Abstracts of Recent Awards (*4*):

- What types of integrations of a conversational interface with media such as text or animations are most effective in providing scaffolding for the learning of physics concepts and for helping students to reflect on their learning?

- To what extent do science course innovations incorporate cognitive science learning theory and principles of effective instruction?
- What are critical factors for the implementation of course innovations?

The NSF Directorate of Undergraduate Education (DUE), under EHR, has different goals and therefore different research questions that it funds. For example, some areas that have been funded in the past (*See "Awards" in 5*) are:

- What is the impact of faculty development workshops on the methods used in the classroom and the effectiveness of these methods?
- What is the impact of innovative modules that introduce real-world questions on the ability of students to identify and recognize the necessity for chemical models?
- What pedagogies are effective means for interactive delivery of scientific visualizations and simulations in guided inquiry software?

A comprehensive list of projects successfully funded by DUE can be found online in their Project Information Research System (PIRS) (*6*).

Other NSF programs, such as Nanoscale Interdisciplinary Research Teams (NIRT) and Nanoscale Exploratory Research (NER), fall outside of EHR and expect investigators to integrate research in a scientific discipline with educational innovation and evaluation. Chapter 13 (*1*) describes how a chemical education researcher can collaborate with scientists on such projects.

The U.S. Department of Education (DOE) is also organized into multiple offices that fund educational research. Science education researchers may find funding from several offices, including the Office of Elementary and Secondary Education (OESE), the Office of Innovation and Improvement (OII), and the Office of Postsecondary Education (OPE). The OII is focused on innovative educational practices, and its Technology Innovation Challenge grants may match the research goals of some chemical education researchers. Projects that have been successfully funded in the past (*7*) include those designed to:

- develop standards-based curricula in a wide range of subjects.
- provide professional development for teachers.
- increase student access to technology and online resources.
- devise techniques for assisting teachers in developing computer-based instruction.
- create strategies for accelerating the academic progress for at-risk children via technology.
- develop new approaches to measuring the impact of educational technology on student learning.

The Fund for Improvement of Post-secondary education (FIPSE), which is sponsored by the OPE, commonly awards comprehensive research grants on curriculum development in science education, including those focused specifically on chemistry. A complete listing of projects funded by FIPSE from 1994 to the present can be found in the FIPSE Grant Database (*8*).

A third federal agency that funds educational research is the National Institute of Health (NIH). The NIH has an office specifically devoted to science education (The Office of Science Education, OSE). This office offers several funding opportunities, including a Science Education Partnership Award (SEPA) that aims to improve scientific literacy. Projects that have been funded through SEPA can be found at their website (*9*).

Other governmental agencies, including the Environmental Protection Agency (EPA), the Office of Naval Research (ONR), the National Oceanic and Atmospheric Administration (NOAA), the National Aeronautics and Space Administration (NASA), and the National Center for Educational Statistics (NCES) also fund projects in science education, and should not be overlooked. A project can sometimes be proposed to multiple agencies or multiple directorates in one agency. For example, NSF has programs that cross between directorates. For submissions to multiple agencies it is important to tailor the proposal to the goals and guidelines of each agency.

While information on all of these previously discussed divisions, directorates, and offices can be found online directly through the websites of the federal agencies through which they are sponsored, it is often difficult for even the most experienced researcher to navigate through and locate pertinent information. Although most agencies offer a link from their homepage directly to the portion of their website devoted to grants, once there, finding a grant that matches your research goals can be fairly difficult. Using the search features of these sites to locate grant opportunities using the keywords "chemical education" may be too specific, and may not return the desired results. For agencies that focus on science, such as the NSF or NIH, searching for grants using the keyword "education" alone may be more successful. Such a search will return information on grants that may not mention chemistry specifically, but which might pertain to research in chemical education. For agencies that focus on education, such as the DOE, searching for the keyword "science" (or even "STEM") will return information on many grants pertaining to science education that may be adapted to fit research in chemical education. Likewise, agencies that fund a wide range of research, such as many of the private agencies listed in Table II, may not yield results on chemical education specifically, but a search for "science education" should reveal more general grants that could be applied to chemical education.

Rather than searching dozens of websites for multiple federal and private agencies that *may* be applicable to your research, it may be less daunting to consult a clearinghouse website. These websites aggregate information on available grants from numerous granting agencies, allowing a single search to reveal multiple grants from a variety of organizations. The website Grants.gov (*10*), for example, consolidates information about grants from multiple government agencies, and even allows electronic submissions of grant proposals. The Foundation Center (*11*) is a parallel clearinghouse for information about private foundations. In addition to their website, one can also subscribe to its newsletter, which delivers proposal requests directly to the researcher weekly via email. The American Association for the Advancement of Science sponsors the Grantsnet Website that distributes information about scientific grants, including grants for science education (*12*). If the planned research involves pre-college education, potential funding sources are also available at SchoolGrants (*13*). Many universities also subscribe to The Grant Advisor (http://www.grantadvisor.com/), an information service that lists grant and fellowship opportunities for both federal and private agencies. The Grant Advisor allows you to search by funding agency, keywords, and academic division (e.g. "education"). Another way to learn what is available in both the private and public agencies is to subscribe to the Community of Science (*14*). Community of Science (COS) is a global resource of information regarding funding for research and other projects across all disciplines. COS aggregates information on available foundation and agency funding, indexed by keywords. A subscriber can select areas of interest and receive a list of the requests for proposals targeted to those areas.

Writing the Proposal

Preparing to Write the Proposal

Those who have never written a grant proposal before should inquire whether their campus grants and contracts office has samples that can be reviewed. Some universities offer grant-writing workshops, short courses, or even provide internal planning grants. These internal planning grants can help researchers plan external research grants by providing release time or summer salary. They also allow the collection of some preliminary data that can be used to enhance an external grant proposal.

The National Science Foundation (NSF), the U.S. Department of Education (DOE) and the National Institutes for Health (NIH) offer a wide variety of resources for grant writing. NSF provides a grant proposal preparation guide (*15*) and lists the criteria for each program on its Website (*16*). In addition, abstracts for each funded project can be viewed at this website. The Department of Education provides a searchable database of program information (*17*) and information on how to write successful grant proposals (*18*). Information on NIH resources for new investigators as well as the NIH criteria for grant reviews can be found online (*19, 20*). In addition to providing information on funding agencies, the Foundation Center also offers a free proposal writing course through their website (*21*).

A good strategy for preparing to write grant proposals is to volunteer to review them. Serving on grant review panels can provide insights into the critical elements that are necessary for a successful proposal. To become a reviewer, researchers can contact a program officer at the funding organization to offer their services. It will be necessary to describe one's areas of expertise and probably complete an application form. Often it is possible to speak with program officers at professional conferences. They can also be contacted by telephone or email. Contact information for program officers is usually posted on the funding organization's website.

Establishing a Timeline

Once the appropriate funding agency and program have been identified, it is important to note the deadline for submission to the funding agency, and to establish a timeline for preparing the proposal. For this timeline, the researcher's institutional deadlines for submitting proposals must be considered—some institutions want the final version of the proposal one week or more before the grant agency's due date. Each campus has various policies that need to be followed as well as additional forms that need to be completed and probably signed by the department chair, dean and other administrators. The university grants and contracts office can provide this routing information along with human subject protocols that might be required for the research. Because many programs within agencies or foundations accept only one proposal from a given institution, it is important that the grants office be alerted to the proposed research at a very early stage.

Some funding agencies, such as some NSF programs, require a letter of intent several months before submission of the full proposal. The letter of intent outlines the overall research plan and allows the program officers to provide feedback that may lead to a better proposal.

Writing the Narrative

With the timeline in mind, the advertisement for the grant (sometimes called a "Request for Proposal" (RFP), "Program Solicitation," or "Call for Proposals") should be carefully studied. A close alignment between the proposed project and the RFP is necessary if the proposal is to be successful. When writing the proposal, one should keep in mind that reviewers will be looking for specific features such as the significance of the project, its approach, feasibility and impact. For the significance of the project, they will want to know the extent to which the project, if successfully carried out, will make an original, important and novel contribution to the field of chemical education and beyond. They will consider the approach and strategy used to describe the conceptual framework, design, methods, analyses, and assessment. Reviewers should clearly recognize a properly developed, well-integrated project that is aligned with the request for the proposal. Literature references are usually required. A realistic timeframe often determines the feasibility of the project, as well as the researcher's documented experience and capabilities, past progress to date, preliminary data, and requested and available resources. Matching university funds or availability of key resources and/or equipment from the proposing institution can often influence the feasibility of a project. To determine the impact of the project, reviewers often look for sustainability when the funding period ends. They also look for transferability of findings to other institutions, especially if different student demographics are involved. The impact of the proposed project will be judged on how the proposal defines success of the project within the given budget of the project.

As an example, the National Science Foundation has a set of "merit review criteria" that every proposal must meet (22):

- Intellectual merit
- Broader impact

The proposal abstract must indicate how the proposed project meets these criteria. The specific criteria on which the proposal will be evaluated typically include the following:

- **Significance:** Extent to which the project, if successfully carried out, will make an original, important, and novel contribution to the field of study;
- **Approach:** Extent to which the conceptual framework, design, methods, and analyses are properly developed, well integrated, and appropriate to the aims of the project (appropriate references are required);
- **Feasibility**: The likelihood that the proposed work can be accomplished by the investigator within a reasonable timeframe, given his or her documented experience and expertise, past progress, preliminary data, requested or available resources.
- **Impact**: The likely impact of the award on the teaching of chemistry and other sciences.

As another example, the Camille and Henry Dreyfus Foundation requires that proposals provide a realistic and detailed description of how a proposed project will impact the field of chemical sciences (23). This foundation also considers the assessment component of the project, especially the criteria that will determine whether a project will be effective or not. Matching funds or institutional support such as reduced teaching load, materials/equipment, and/or student assistant support are also required.

Establishing the Budget

A significant feature of the grant proposal is the outline of the projected budget for the project. The budget should provide a realistic level of funding for the project and meet the funding agency's budget criteria. As an example using NSF budget guidelines (15), allowable expenses include graduate assistant or undergraduate assistant stipends, summer salaries for the senior personnel, a small amount of release time, supplies, fees for services such as transcription, computer programming, lab analyses, participant stipends, consultants, printing costs, and travel to scientific meetings for dissemination or collaboration. The program director should be consulted to find out if an external evaluator is required. On a research grant, the role of the external evaluator is to provide oversight and advice while checking that the project is progressing and that the goals are being met. If an external evaluator is required, the budget should include funds for the evaluator's time and travel.

For salary and stipends, the percent effort or months per year should be included. Fringe benefits will need to be included and can change annually. The university research office can assist in these calculations as well as in estimating projected increases for a multi-year grant proposal. With respect to consumable supplies, typically chemical education projects may require questionnaires, brochures, folders, papers, pens, laboratory supplies or equipment as well as office and computer supplies. Some other expenses may include recruitment costs such as advertising or small incentives for participants, production of marketing materials, website development, photocopying, costs of reprints and publishing, participant meals and lodging, as well as audiovisual development. All these concerns should be taken into account as the proposal and budget are developed.

If the proposal needs an institutional match, a meeting with the department chair, dean and perhaps the provost should be set up before submitting the proposal. The researcher will need to clarify precisely what is expected from the grant agency for this match, whether it be tuition assistance, administrative support, space and/or reduced teaching load. Even if the RFP does not specify an institutional match, it can be helpful to obtain some matching funds or equipment.

Specific questions regarding the proposal's criteria and/or budget can be discussed directly with the program officer. The program officers are usually happy to provide assistance via e-mail or telephone.

The university will add a certain percentage of overhead expenses (also referred to as indirect costs) that must also be included in the budget. Overhead expenditures generally include operating costs such as electricity, water, gas and grant administration costs. A funding agency can specify a specific limit on overhead costs, regardless of the institution's policy. Either way, it will be the responsibility of the researcher to construct a realistic budget that satisfies both the institution's criteria and the funding agency's criteria.

Besides providing a budget, most grant agencies require a budget justification. Sample examples of how to justify positions, supplies and travel are found in Appendix A.

Incorporating Collaborations

Some grants are collaborative in nature and different models can be followed, particularly if there is a chemical education research component on another type of project. Projects for which the chemical education researcher is a subcontractor on a larger grant are described in more detail in Chapter 13 (*1*). If the chemical education researcher is the PI and the proposal involves collaborators from the same or other institutions, each collaborator will need to provide a letter specifically describing their roles in the project and willingness to participate as well as a curriculum vitae. If the collaborators are from different educational institutions and will receive compensation from the proposed project, a subcontract between the originating institution and the collaborating institutions must be prepared. Arranging a subcontract can take a few weeks, so it is important to budget appropriate time for this activity. A collaborator with significant responsibilities might serve in the role of a Co-Principal Investigator, whereas one who is spending little time on the project may be better included in the proposal as a consultant. In this case, a subcontract may not be necessary.

Finalizing the Proposal

Once a draft of the proposal has been completed, it can be most helpful to ask a trusted colleague who is familiar with the grant program to review the proposal. If suggested reviewers are requested, it is wise to suggest reviewers who are familiar with the research area proposed, especially if it is an unusual one. Before submitting the proposal it is also a good idea to set up a checklist of items based on the Request for Proposals and to review the list to make sure that all the items have been addressed.

It is important to verify that the proposal follows the principles discussed in this book and, for federal proposals, those in *Scientific Research in Education*

(*24*). In addition, the NSF Strategic Plan (*25*) can be consulted to ensure that the proposal is in line with current NSF strategies. Common faults in proposals for research in science education include failing to state the research questions explicitly, proposing research questions that are not testable, failing to show how the research questions arise from the literature, and stating research questions that are trivial. Review panels generally want proposals to address all the disciplinary perspectives and it is important for panels to easily find the overarching logic of the proposal. The proposal should be completed and submitted before the published deadline. Such deadlines are firm deadlines and no exceptions are made.

Acknowledgement

The authors of this chapter would like to acknowledge helpful conversations with Elizabeth VanderPutten and Gregg Solomon, Directorate for Education and Human Resources, National Science Foundation.

References

1. Zare, R. N., Questions to Chemical Educators from the Chemistry Community. In *Nuts and bolts of chemical education research*, Bunce, D.; Cole, R., Eds. American Chemical Society Symposium Series: Washington, DC, 2007.
2. Missouri State University. Office of Sponsored Research and Programs. http://www.srp.missouristate.edu (January 30, 2007).
3. Sawrey, B. A., Collaborative projects: Being the chemical education resource. In *Nuts and Bolts of Chemical Education Research*, Bunce, D.; Cole, R., Eds. American Chemical Society Symposium Series: Washington, DC, 2007.
4. National Science Foundation. National Science Foundation Research and Evaluation on Education in Science and Engineering (REESE) Program. http://www.nsf.gov/funding/pgm_summ.jsp?pims_id=13667&org=EHR&from=home (October 20, 2006).
5. National Science Foundation. Division of Undergraduate Education (DUE) Home Page. http://www.nsf.gov/div/index.jsp?div=DUE (January 30, 2007).
6. National Science Foundation. NSF/DUE Project Information Resource System (PIRS). https://www.ehr.nsf.gov/pirs_prs_web/search (January 30, 2007).
7. U.S. Department of Education. Technology Innovation Challenge Grant Program. http://www.ed.gov/programs/techinnov/index.html (January 30, 2007).

8. U.S. Department of Education Office of Postsecondary Education. FIPSE grant database. http://www.fipse.aed.org (January 30, 2007).
9. Science Education Partnership Award Program. SEPA Programs Listed by Funding Year. http://www.ncrrsepa.org/program/Year.htm (January 30, 2007).
10. U.S. Department of Health and Human Services. Grants.gov Home Page. http://www.grants.gov (January 30, 2007).
11. Foundation Center. The Foundation Center Home Page. http://www.fdncenter.org (January 30, 2007),
12. American Association for the Advancement of Science. ScienceCareers.org Funding. http://sciencecareers.sciencemag.org/funding (January 30, 2007).
13. Fernandez, D. SchoolGrants Home Page. http://www.schoolgrants.org (January 30, 2007),
14. CSA. Community of Science Home Page. www.cos.com (January 30, 2007).
15. National Science Foundation Office of Budget Finance & Award Management. NSF 04-23 Grant Proposal Guide. http://www.nsf.gov/pubs/gpg/nsf04_23/ (October 22, 2006).
16. National Science Foundation. US NSF Home Page. http://www.nsf.gov (January 30, 2007).
17. U.S. Department of Education. U.S. Department of Education Programs and Resources. http://web99.ed.gov/GTEP/Program2.nsf (January 30, 2007).
18. U.S. Department of Education. Grantmaking at ED Table of Contents. http://www.ed.gov/fund/grant/about/grantmaking/index.html (January 30, 2007).
19. National Institute of Health. New Investigators Program. http://grants1.nih.gov/grants/new_investigators/index.htm (January 30, 2007).
20. National Institute of Health. NOT-OD-05-002: NIH Announces Updated Criteria for Evaluating Research Grant Applications. http://grants.nih.gov/grants/guide/notice-files/NOT-OD-05-002.html (January 30, 2007).
21. Center., F. Tutorials: Proposal Writing Short Course. http://foundationcenter.org/getstarted/tutorials/shortcourse/index.html (January 30, 2007).
22. National Science Foundation. Instructions for Proposal Review. http://www.nsf.gov/pubs/1997/iin121/od9708a.htm (January 30, 2007).
23. The Camille & Henry Dreyfus Foundation. Special Grant Program in the Chemical Sciences. http://www.dreyfus.org/sg.shtml (January 30, 2007).
24. Shavelson, R. J.; Towne, L., *Scientific research in education*. National Academy Press: Washington, DC, 2002.
25. National Science Foundation. National Science Foundation Investing in America's Future Strategic Plan FY 2006-2011. http://www.nsf.gov/pubs/2006/nsf0648/nsf0648.jsp (October 22).

Appendix A: Sample budget justifications

Personnel Costs: Salaries for senior personnel: Include percent effort or months per year. Don't forget to include fringe benefits for senior personnel. Fringe benefits are identified as a separate item in a budget. Because fringe benefit rates change annually, the rates used should be checked with the institutional research office. In general, reviewers do not look favorably on proposals for which the largest amount requested is for the Principal Investigator's salary. Participation in research is felt to be an expected part of the duties of a researcher. However, a month of summer funding is not only expected, but also desirable, as it shows that the PI is setting aside time to devote to the project. Note that time needs to be given in academic year months for academics. The following are sample descriptions of budget justifications for project personnel:

"The Project Coordinator will facilitate the research by making arrangements for the installation of the software, coordinating communications, meetings, and research protocols among the various campuses, preparing research instruments, and participating in transcription and data analysis."

Salaries and tuition for GAs: Graduate assistants may not have fringe benefits or may have a different rate, but they do have tuition costs, which should be itemized.

"Two graduate assistants will recruit research subjects, collect research data and analyze it. They will also help to disseminate project findings through presentations at professional meetings."

Travel: Provide a justification for each trip. List destinations of the travel, number of trips planned, who will be making each trip, and approximate dates. If mileage is to be paid, provide the number of miles and the cost per mile.

"The PI, Co-PI, and one graduate student will travel to two professional meetings per year, where they will present results and participate in planning sessions with project personnel from other institutions. Estimated travel costs per person per 3-4 day trip are as follows:

- Airfare: $400
- Hotel: $600
- Meals: $200
- Taxi and miscellaneous: $75
- Total per person per trip: $1275 @ 2 trips/year = $3825/year"

Supplies: List each item requested (unit cost/item and quantity). Relate each item to specific program objectives such as:

"Materials and supplies required for the project include digital videocassettes and audio tapes for the qualitative studies, transcription equipment, diskettes, and graphics software. Funds are included for the printing of the tests to be used in the quantitative studies, and for posters and materials to be distributed at presentations."

Other costs: Include the cost of teleconferences, incentives for research participants, the production of marketing materials, development of web site, photocopying, etc., the costs of reprints and publishing, transcription costs, and fees for computer programmers or designers.

Chapter 4

Constructing Good and Researchable Questions

Diane M. Bunce

Chemistry Department, The Catholic University of America, Washington, DC 20064

No matter who generates the question to be investigated (chemistry community, granting agency, or researcher), it should be constructed in such a way that it is researchable. There are five components to constructing a good researchable question. They are: 1) Is the question worth asking? 2) Is it feasible? 3) Who will be studied? 4) How will it be investigated? and 5) What is the potential "take home" message (result) of the investigation? A careful examination of each of these points will help with the development of an important, realistic question, or set of questions, that can be rigorously investigated. Asking the question should lead to an investigation that can provide convincing results that speak to the researcher or the community that raised the question in the first place. This chapter will take the reader step-by-step through the process of identifying a problem and developing the resulting question to be investigated. Such a process will determine certain parameters for how the investigation will be conducted, the data collected, and the conclusions drawn.

Introduction

Investigating questions of teaching and learning is no longer the sole responsibility of people specifically trained in the theories and practices of chemical education research. More and more, universities, departments, funding agencies, and professors are asking questions of how students learn and how a particular teaching method will affect students. In order to engage in thoughtful and valid investigations of such topics, many people find that they need more direction on how to frame the inquiry. As in all good investigations, the place to start is with a good question. As easy as this sounds, devising a question that can be rigorously investigated and answered convincingly can be a stumbling block for even the most experienced researcher. The question must be important enough to warrant the time and energy it will take to investigate it, but also practical enough to make the study feasible given the resources and tools available. One thing that is often overlooked is that the question asked should determine, at least partially, both the general research design and the instruments needed to investigate it. So it is especially important to realize that *how* the questions are asked will directly affect the type(s) of investigation conducted. Before beginning the investigation, the question should be examined to determine if it will address the central issue that is driving the investigation. If not, then the question should be reconfigured to more adequately address the issue.

Components of a Good Question

There are five components of a good question, which are often addressed in a nonlinear fashion. The first component involves identifying the problem. What is it you want to know? Putting this goal in your own words first, will help you move quickly to the heart of the matter. For example, do you want to know why students have trouble learning chemistry?

This problem must be further refined to include what you specifically want to know about the problem. In other words, do you want to know why students have trouble leaning chemistry in general or do you want to know why the freshmen enrolled in a first semester general chemistry course fail or withdraw from the course at such a high rate? In this example, the first question about why students have trouble learning chemistry is a life's work and involves identifying and measuring/controlling a large number of variables. The second question is more limited in scope and thus more practical in terms of experimental design. Moving from a global problem to the situation that you really want to investigate is an important process to engage in. This process will provide you with a more limited and feasible question to address which makes the design of a research plan more realistic. Of course, moving from the general topic of how students learn chemistry to your specific situation of freshmen dropping or failing general chemistry is a big step and takes some time and reflection on your part. You can move through the process successfully by asking yourself what it is *you* really want to know.

The second component of designing a good question is whether it is feasible to study. For instance, you might want to know if all freshmen suffer the same withdrawal and failure rate in general chemistry, but the only group that you have direct access to are the students enrolled in your university. Therefore, the question you investigate has to be limited to "Why do students enrolled in first semester general chemistry at University X experience such high withdrawal and failure rates?" The conditions you have placed on your question help make the investigation feasible in terms of time, energy and money, but they also may make the question so specific to your institution that the results may be of limited importance to those outside your institution. A point to consider when designing a question is whether you want to investigate your specific situation or do you want to publish the research to add to the knowledge of the larger research community. If your goal is the former, then investigating a small, focused question of importance to the local community is appropriate. However, if you wish to ask this question and publish the results to inform the larger community then it will be necessary to broaden the question and link it to a bigger question of importance to learning. Usually generalizability is a result of a quantitative design where the statistical considerations of sample size, significance and power are addressed.

The third component of the question-writing process, dealing with whom you will study, is often defined by the time you get to this step. In the case we are discussing, investigation of the withdrawal and failure rates had already been confined to freshmen, rather than other classes; those enrolled in first semester general chemistry, rather than in other courses; and those students meeting these two criteria who were enrolled at University X, as opposed to students enrolled in all colleges in the country. If these things had not already been specified, now would be the time to make them explicit.

The fourth component of question writing is measuring what you want to know. Addressing this issue will force you to confront what it is you *really* want to know. In the case of our example, if you want to know why freshmen experience such high withdrawal and failure rates, you should ask *them*. It doesn't pay to speculate on the reasons. The purpose of the investigation should be to find out why they withdraw or fail freshman chemistry. There are two ways to investigate the reasons for failure and withdrawal from a student's point of view. The first is to interview those students individually who are withdrawing and failing general chemistry. The second is to develop a survey that can be used effectively with a larger number of students and that does not have to be administered face-to-face with the student. Since the important variables in this situation are not well known, it is wise to interview a limited number of students first and then use the information gleaned from these interviews to develop the survey questions. This type of qualitative research design is implicit in the question that is being developed. We are not asking if one method of intervention with freshmen in this situation is better than another. We are asking why this situation of withdrawing from or failing freshman chemistry happens. Questions of *why* or *how* usually indicate that the pertinent variables responsible for what we see are not well known. The purpose of the research then, is to determine what the variables are. This type of qualitative study, identifying the pertinent variables, can be followed by a quantitative study investigating whether one type of intervention is more effective than another at reducing the number of withdrawals and failures in freshman chemistry classes at University X.

The fifth component of writing a good question is to make sure that you have a clear "take home" message in mind. In other words, what do you hope to be able to say at the end of your investigation? For instance, in our example, we might want to be able to identify three or four critical areas where intervention may affect the number of freshmen withdrawing or failing general chemistry at University X. If the "take home" message you identify addresses your concerns, then you can be confident that the question you are asking is the one you are really interested in. If the "take home" message that your research question leads to is not what you are interested in, then now is the time to re-evaluate your question and modify it so that you *are* satisfied with its potential "take home" message. The next part of the chapter will look at these question components in more detail.

1. Identifying a problem

If you have been given a problem to investigate by your administration, department, or funding agency, then you already have a starting point. If not, then the goal is to identify a problem you want to investigate

One way to generate problems to investigate is by brainstorming. Maykrut and Moorehouse (*1*) recommend thinking about the ideas that you are interested in and which aspect of the topic you would like to investigate. Then, their advice is to pick one idea, write down anything related to it, look for patterns in what you have written, connect related ideas, and draw a concept map relating the main and secondary ideas.

For instance, if I am interested in the implementation of teaching innovations, I could consider several NSF-supported teaching innovations such as Processed-Orientated Guided-Inquiry Learning (POGIL) (*2*), Peer-Led Team Learning (PLTL) (*3*), use of ConcepTests (*4*) in lecture, or the use of technology in teaching. Having brainstormed these ideas, I could then write down things that are related to my choice. For instance, if I chose the use of technology in teaching as the main idea, the list of things related to it might include computers, tablet PCs, web-based information, WebCT or Blackboard, online quizzing, online discussions, electronic office hours, and/or student response systems (clickers). Reviewing these ideas for a pattern, I might find that ways to "use technology to engage students in the learning process" is a common theme. This would be followed by my constructing a concept map relating the broad and specific ideas on the topic. When I have identified the topic, it is wise to spend some time evaluating the importance of asking a question about it. Is the topic important to the community-at-large's understanding of a bigger phenomenon or is the topic important only to the local community? How you answer this question will tell you whether the research is likely to be publishable.

2. Feasibility

In order to study the effect of technology on student learning, you must have access to students and professors who are actually using the form of technology you are interested in studying. The people involved in your study should reflect all the pertinent stakeholder groups. In this example, it would be important to include both students and professors since both are involved in using the technology for the course. In some teaching situations, teaching assistants would represent another group of stakeholders and their input should be sought at well. Gaining access to the people is the first part of this analysis, but it is not enough. If you want to know how often and at what times students use these forms of technology, then you can probably gain your information unobtrusively by embedding counters in the software programs that the hardware uses. However, if you want to know the effect that using these forms of technology has on student learning, then you will have to plan more intrusive ways of collecting data.

In either case, unobtrusive or intrusive, you will need approval to collect data on this topic from your school's Institutional Review Board (IRB). Each institution that accepts federal research money is obligated to set up an Institutional Review Board to review all research dealing with human subjects. It is wise to contact the IRB at your institution and become familiar with their criteria for such research before proceeding very far with your research question(s). Approval can range from an exemption (the research is within the parameters of normal evaluation of classroom teaching) to a full review (submission of a detailed description of the questions asked, tools used, inconvenience/discomfort caused to subjects, and a copy of the permission sheet that subjects will be asked to sign before data can be collected).

When determining the feasibility of a study, you should think about access, amount of time, energy and money required on both your part and that of the subjects, and whether you can collect the type of data that will address the question you want to ask. In this current example, it might be necessary to access the computer programs that are used in these technologies, develop questions that measure student understanding to be used in conjunction with these technologies, and collect a copy of the students' responses to these questions either through the technologies themselves, interviews with students, or copies of student test papers given throughout the semester. If one were interested in student-use patterns or perceptions of technology, the data collected might include counters embedded within the software, observations, interviews, and/or surveys.

3. Subjects for Study

The subjects included in the study should be appropriate for the question you are asking. For instance, if you want to know what students gain from the use of technology, it is important to study students who have access to the technology in question. If, on the other hand, you want to study the withdrawal and failure rates of freshmen in general chemistry, studying only nonscience majors or students in an advanced general chemistry course may not be the only subjects you want to investigate. The best route is to include all the students who affect or are affected by the phenomenon under investigation. In our example, that might be the nonscience majors, advanced general chemistry students, and the students enrolled in the main general chemistry course. The population you study should be the population most directly affected by, or involved in, the question you raise. They must also be a population you can gain access to for the measurements you want to make. Studying a nonscience majors' introductory chemistry course will probably not yield the same results that a general chemistry course for chemistry/other science majors would on the question of failure/dropout rates. But if you don't have access to the general chemistry science majors and you do have access to the nonscience majors, then perhaps you should modify your question to reflect the problems experienced by this population.

4. Measuring variables

One of the classic problems encountered in this type of research is that the tools used don't measure the variables in the question asked. Instead, tools are often used that measure convenient variables even though these variables don't address the question. For instance, if you want to know if a particular teaching innovation results in student learning, using a survey that asks students if they like the innovation does not necessarily answer the question of whether the innovation increased student learning. It is true that a positive student perception of the innovation can result in more time on task and more incentive to learn, but this is just one part of learning effectiveness. Too often student attitude towards a teaching innovation is substituted for evidence that enhanced learning has taken place. This happens because student opinion is easier to measure than student learning. To be effective, the tools used must produce data that are directly related to the question(s) asked.

If at all possible, using a proven instrument to measure a variable is preferable to devising a new instrument that then must be tested for reliability and validity before it can be used to collect data. However, in many cases, instruments do not exist for the particular question being investigated. They must be constructed and that takes time. When developing instruments it is important to remember that the focus and precision of the instrument used to measure the variables of your question must match the question you want to investigate. For instance in a study that involved the effect of ConcepTests delivered via clickers on student conceptual understanding (*5*), the tools consisted of an electronic copy of the students' responses to ConcepTest questions. This use of the clicker software provided an opportunity to collect information on student achievement in an unobtrusive manner. However, it did not fully address the question of whether students' conceptual understanding had changed as a result of answering ConcepTests. To investigate this part of the question, an intrusive approach of having students fill out index cards with the *reason* for the answer they chose was employed. Together these two tools helped address the question of whether student conceptual understanding was developing with continued use of ConcepTests. A good source of simple research instruments, such as the index card system used in this example, can be found in a book on classroom achievement techniques by Angelo and Cross (*6*).

5. Take Home Message

When the question is investigated, it is important to clearly state the outcome. In other words, what do you want to be able to say as a result of your investigation? This "take home" message is the real reason you asked the question in the first place. It is important at this stage in the question development process that you re-examine if the question you asked has the possibility of generating a take home message you are interested in. If not, then it is time to revise your question.

Using one of the examples we have discussed in this chapter, a take home message for the technology question might be that by using technology (clickers), we can engage students in the learning process and measure their progress towards a deeper conceptual understanding of chemistry concepts over a given period of time. For the withdrawal and failure question, our take home message might be that if students can relate to a given individual in the chemistry department on a personal level, and if they receive the help they need when they need it, the withdrawal and failure rate for freshmen in general chemistry is significantly reduced. Of course, our investigation may not result in these specific take home messages, but if the study has the potential to deliver these results and if these results correlate to the question we want to ask, then we know that we have asked a good question.

Importance of Theory-Based Questions

Chemistry research questions are usually based upon a chemical theoretical foundation. The same should be true of chemical education research questions. If chemical education research questions are asked without connecting them to overarching theories of cognitive psychology, sociology, information sciences, visualization, learning, outreach, leadership, etc., then the questions addressed will have little impact on our overall understanding. They will not extend our knowledge of how learning takes place. It is hard to think of the Human Genome Project accomplishing all that it has without the underlying model of how DNA operates. Without the unifying principle of DNA, the individual discoveries from countless labs around the world would be scattered bits of knowledge with little impact on our understanding. But with the theory of DNA and a concerted effort to ask questions that fit that theory, we now have integrated knowledge that greatly expands our vision of what is possible. So too, the questions of chemical education research must extend theories of learning, teaching, and outreach in order to affect our knowledge of teaching and learning in a meaningful way (7).

There are theories of learning that many people mention in passing when designing a study. Popular among these are the theories of Piaget and Constructivism, but mere hand waving without a true understanding of what the theory is saying is not adequate. People who ask questions in the chemical education research domain have the same responsibility chemists have for both searching the literature for theories that form the basis for the question asked and for understanding the theory thoroughly enough to accurately use it to help interpret their results. Mike Abraham's chapter (8) in this book will look at the question of theory-based research in more detail

Mechanics of Writing Questions

Writing good questions involves the principles of identifying a problem within a relevant theoretical framework; becoming familiar with the research that has already been done on the subject; identifying who, what, when, where and how (9); and testing that the question asked matches the outcome that can be produced through the methods chosen to investigate it. In this section, we will concentrate on the who, what, when, where, and how of question writing.

Subject and Verb

To answer *who* the research is directed at, we must specify our population. Is our population all students in chemistry, only those taking general chemistry, or the subgroup of this population who are freshmen taking first semester general chemistry? The *when* might be during the upcoming semester (Fall, Spring or Summer). The *where* might be at University X or community colleges within the state consortium or even those students at my university who are taking general chemistry at 9AM. The *what* could be their conceptual understanding of the equilibrium concept, their achievement after having used the POGIL approach, or their conceptual understanding developed when engaging in ConcepTests. The *how* deals with the experimental design that will be used to collect data related to the question asked.

After addressing these aspects of the question, we are still left with the mechanics of writing the actual question. These mechanics involve specifying a noun (who) and a verb (what) along with modifiers (when, where, and how). It is the verb in our research questions that requires the most attention at this point. It is not enough to ask if a certain intervention will "increase learning". The term "increase learning" must be made operational or testable. Learning can mean different things to different people. To some, learning means being able to provide chemical definitions word-for-word from the text or notes; or completing a mathematical chemistry problem just like the one shown in class, but this time with different numbers. For others, learning means the ability to transfer knowledge from a problem they have seen before to a novel problem. It also can mean being able to think creatively and apply previously learned concepts to a new situation. For still others, it means being able to go beyond the correct answer to the reason *why* something happens chemically. As a result, using a term "increase learning" in a question needs to be operationalized so that people reading the question understand exactly what you mean by the term.

Even if learning is replaced with the word "achievement" in a question, misunderstandings can still exist. If your question is whether students will "achieve more" after experiencing one type of instruction or another, you must define this term carefully. Does "achieve more" mean attain higher test scores? And if so, then on what kind of test—standardized tests like the ACS tests or nonstandardized teacher-written tests? What kind of questions are on these tests—memorization, algorithmic, or conceptual understanding questions? Does "achieve more" mean statistically higher scores than students who do not experience the same type of instruction? If not statistically higher, then there is a chance than any change is simply due to chance and not reproducible.

Making the verb in your question specific and defining it operationally will also help in the design of the study. With an operationally-defined verb, you will be better able to describe the types of tools used and the data collected as you investigate your question.

Intervening or Confounding Variables

When your question is formulated, it is important to be aware of intervening or confounding variables that could interfere with the investigation of your question and/or interpretation of data. It is worthwhile making a list of all of the things that could interfere with your results because these could affect how the question is asked or investigated. For instance, if you choose to explore the effects of a particular teaching innovation in a general chemistry class compared to a class that is not using the innovation, you might consider the students in each class on one or more pertinent variables. For instance, sometimes due to scheduling, classes may have a higher than normal distribution of engineering or nursing majors. If your innovation is mathematically based, then using it with the class having a large percentage of engineering majors may result in false positive results. Likewise if the mathematically-based innovation is introduced in a class that has a large distribution of nursing majors, you may experience a false negative. To handle this situation, a measure of math aptitude could be used to verify and statistically control for a significant difference in mathematical aptitude between two classes. This identification of an important intervening or confounding variable and attempt to control it either through the experimental design or statistical methods is vital for a correct interpretation of your results and thus a genuine answer to your question.

When intervening variables are identified, they can also be analyzed through sub-questions. For instance, when investigating whether one approach is better overall than another, sub-questions could explore this effect for women vs. men, high math-achieving vs. low math-achieving students, or traditional vs. returning students.

Relationship between Question Asked and Experimental Method

The type of question you ask has direct bearing on the choice of the most appropriate research design. For instance, questions that ask *how* or *why* are best answered by qualitative methods. Questions that address *differences between groups* are best answered by quantitative methods. Some questions contain aspects that include both how/why and differences between groups. These questions can be addressed by a mixed methods design. A book by Creswell (*9*), the NSF publication on mixed methods (*10*) and the chapters in this book on qualitative (*11*), quantitative (*12*) and mixed methods (*13*) are good places to start learning about the differences among these three approaches.

Final Question

It is important to remember that the question is never really in final form until the research is completed. The question is a living document that guides the investigation and is constantly being modified by what can and is being measured. The question requires operational definitions to fully explain what is being asked and takes into account intervening variables that could affect the correct interpretation of results. In reality, the question is both the first and the last thing written for an investigation. It defines the research design and the tools used for collecting data and in the end, it is the most important part of the investigation.

Some authors (*14*) offer checklists for writing good research questions. These lists include suggestions such as 1) The problem should be explicit and able to be investigated in a meaningful way 2) There are no ambiguous terms in the question 3) Underlying assumptions are made explicit 4) Intervening variables are controlled 5) Research methods and analyses have been addressed 6) An appropriate theory has been selected 7) Applicable literature has been surveyed and 8) The general type of experimental design (Qualitative, Quantitative or Mixed Methods) has been chosen. These points have all been addressed in this chapter.

Conclusion

Chemical education research, like all research, chips away at big questions (ex. how do students learn) by providing well-documented attempts to isolate and investigate the effects of many pieces that form part of the bigger questions. Attempts to address a multi-variable problem like learning with a single experiment are doomed to failure. In chemical education research, like all science, all the pertinent variables have not yet been identified and all the tools needed to measure these variables do not yet exist. This is the basis of scientific investigation—to ask new questions, or old questions in new ways, and devise a way to investigate them. The most important part of the process is constructing good questions. Using inappropriate or vague questions to drive the research is one of the main reasons why many educational experiments in the past have shown no significant results. Chemical education research must be carried out with the same standards of excellence and the same attention to detail that research in other areas of science use and that process starts with asking good questions!

References

1. Maykut, P.; Morehouse, R. *Beginning Qualitative Research: A Philosophic and Practical Guide*; The Falmer Press: Washington, DC, 1994.

2. POGIL: Process Orientated Guided Inquiry Learning. http://www.pogil.org (accessed Dec 2006).
3. Gosser, D. K.; Cracolice, M. S.; Kampmeier, J. A.; Roth, V.; Strozak, V. S.; Varma-Nelson, P. *Peer-Led Team Teaching: A Guidebook*; Prentice-Hall, Inc.: Upper Saddle Brook, NJ, 2001.
4 Landis, C. R.; Ellis, A. B.; Lisensky, C. G.; Lorenz, J. K.; Meeker, K.; Wamser, C. C. *Chemistry ConcepTests: A Pathway to Interactive Classrooms*; Prentice-Hall, Inc.: Upper Saddle Brook, NJ, 2001.
5. Bunce, D. M.; VandenPlas, J.; Havanki, K. *J. of Chem. Educ.* **2006**, *83*, 488-493.
6. Angelo, T. A.; Cross, K. P. *Classroom Assessment Techniques: A Handbook for College Teachers*, 2nd ed.; Jossey-Bass: San Francisco, CA, 1993.
7. Leonard, W. H. *J. College Science Teaching* **1993**, *23*, 76-78.
8. Abraham, M. R. In *Nuts and Bolts of Chemical Education Research*; Bunce, D. M., Cole, R., Eds.; American Chemical Society: Washington, DC, 2007.
9. Booth, W. C.; Colomb, G. G.; Williams, J. M. *The Craft of Research*; University of Chicago Press: Chicago, IL, 2003.
10. Creswell, J. W. *Research Design: Qualitative, Quantitative, and Mixed Methods Approaches*, 2nd ed.; SAGE: Thousand Oaks, CA 2003.
11. Frechtling, J.; Westat, L. S. *User-Friendly Handbook for Mixed Method Evaluations*; National Science Foundation Directorate for Education and Human Resources, Division of Research, Evaluation and Communication: Arlington, VA, 1997.
12. Bretz, S. L. In *Nuts and Bolts of Chemical Education Research*; Bunce, D. M., Cole, R., Eds.; American Chemical Society: Washington, DC, 2007.
13. Sanger, M. In *Nuts and Bolts of Chemical Education Research*; Bunce, D. M., Cole, R., Eds.; American Chemical Society: Washington, DC, 2007.
14. Towns, M. H. In *Nuts and Bolts of Chemical Education Research*; Bunce, D. M., Cole, R., Eds.; American Chemical Society: Washington, DC, 2007.
15. Issac, S.; Michael, W. B. *Handbook in Research and Evaluation for Education and the Behavioral Sciences*; EdITS: San Diego, CA, 1980.

Chapter 5

Importance of a Theoretical Framework for Research

Michael R. Abraham

Department of Chemistry and Biochemistry, University of Oklahoma, Norman, OK 73019

Research in chemical education should be theory-based. Although this assertion would seem to most to be non-controversial, it is common to find research studies in the literature without the grounding of theory. Some would claim that theory is not necessary, because research should be focused on answering narrow questions that would more likely appeal to practitioners. Nevertheless, theory-driven research has advantages for the development and growth of the discipline of chemical education. Theory can guide research, practice, curriculum development, evaluation, and help develop effective instructional tactics and strategies. Furthermore, it is not clear that theory-free research can exist. A researcher who doesn't articulate an underlying theory about how learning takes place is probably still operating on a theory of learning and making educational decisions about how learning takes place. Theory about learning chemistry can be derived from established disciplines like psychology, sociology, and philosophy. A consideration of prominent learning theories derived from these sources demonstrates the power of theory-based research strategies.

Introduction

Efforts at improving instruction in chemistry have focused on curricular issues, on the development and implementation of instructional materials, and on strategies of instructional presentation. To a large extent, these efforts have not been based on educational research but have reflected beliefs, anecdotal stories, educational folk lore, and/or historical practice. In recent years, there has developed a body of knowledge that is wide-ranging enough to guide instruction in chemistry and that can be used to develop theoretical structures. At the very least, this knowledge can provide useful hypotheses about who, what, when, where, why, and how to teach chemistry that can be tested in instructional settings and that can add to our understanding of teaching and learning.

Research and Theory

The premise of this chapter is that research should inform the practice of chemical education and that research is more useful and powerful if it is theory-based. This position seems self-evident and to need no justification, especially for academics trained in chemistry. However, this premise is not uniformly believed or applied. First of all, there is an argument in some educational circles that education is a discipline for which theory is not necessary or practical. Secondly, practitioners of the educational process, for the most part, do not use and/or are unaware of research findings. As a consequence, their ideas about what constitutes effective practice can be limited or flawed.

The modern history of science education can be characterized as having a concern for making itself legitimate as a discipline by developing a theory-base. Discussions in the 1970's and 80's were concerned with the loose organization of the discipline of science education and the need for organizing paradigms to guide research (*1*). In 1982, a survey of the 35 largest graduate centers for science education asked science education professionals to identify the current problems in the discipline and what was needed to address those problems. The response was that a theory-base for science education was needed to guide research and practice that was reflective of the nature of science, the nature of the learner, and the goals of education (*2*).

Because of the close bond with the natural sciences, science education researchers at first used experimental science as a model for doing research. They weren't alone in this. Many of the social sciences, such as psychology, also were drawn to the legitimacy of scientific research as a standard. Because of this, theory traditionally played a central role in research. In recent years the view that experimental research should be the model for educational research has been questioned (*3*). Experimental research, it is said, is limited by its methods and by the kinds of questions that it addresses. These limitations often lead to trivial findings about issues that are not important to practitioners. A better approach, it is argued, is a naturalistic, observational approach. This challenge of scientific paradigms can lead to an antitheoretical stance (*4*).

The Controversy: Theory vs. Empiricism

There are several counter-arguments to the assertion that research should be theory-based. The first of these comes out of the original acceptance of the experimental sciences as a model for research. Research practices in education, as was true of psychology, were originally modeled on the experimental sciences. But as psychology branched out from modeling behavior with animal studies to the study of humans in social conditions, methodologies expanded to include more descriptive and less experimental methods. These ideas influenced educational research. The observation was made that education was carried out in environments that more closely resembled anthropological field research than laboratory science. The change in approach was reflected in a change in emphasis away from experimental studies to more observational (descriptive - exploratory) studies. (A survey of types of research studies in the *Journal of Research in Science Teaching* from 1980 to the present will show this trend.) An outcome of this shift in emphasis was a shift away from theory. Experimental approaches are more likely to be theory-based. Descriptive approaches tend to be more empirical – that is more focused on answering specific questions or deriving information from observation. Better examples of this type of work do attempt to link to or build theory. Ideally, descriptive approaches can be an exercise in theory seeking, with the consequent follow-up concerned with testing the theory in subsequent experiments. However, often this doesn't happen; the results are just reported, and others are left to try and figure out what to do with them.

The de-emphasis of the experimental sciences as a model for educational research led to the methodological wars between qualitative vs. quantitative research. New researchers were deciding which camp they were in before they even decided which question to ask and if one approach was more suited to the question than another. Method was usurping purpose.

Recently, researchers have taken a more eclectic view of research methods by matching the methods to the research question at hand; taking advantage of the strengths and adjusting for the weaknesses in both qualitative and quantitative approaches (*5*). However, qualitative research methods are not necessarily antithetical to the role that theory can play in research. Indeed, the same qualitative data set can yield different, but not contradictory, conclusions depending on the theory used for its interpretation. Although it might be true that quantitative methods, in general, are based on theory, and qualitative methods are, in general, used to generate theory, the recent emphasis on using mixed methods is more powerful if theory is central to the endeavor in either case. This argument is taken up in more details in chapters 7, 8 and 9 of this volume. *(6, 7, 8)*

Focus on Questions

Another argument against theory in educational research is represented by Westmeyer *(9)*. He questioned the need for theory in science education research because theories could be taken from other disciplines like psychology. He further argued that researchers should just answer questions of interest to practitioners in the discipline. The first argument is not really an argument against theory. It is just an argument that theory in science education research can be derived from other more mature disciplines. However, even if derivative, theory is usually modified, refined, or interpreted when used in a different context. The second argument, that research should be question-based, implies a purely empirical emphasis for research. Geelan *(10)* argues that research in education, in order to be practical for practice, should avoid theory and instead focus on the needs of teachers and students. This might be a response to the concern on the part of researchers that practitioners don't seem to be utilizing the results of research. Practitioners seem to just want to be told what works *(11)*. A possible source of this attitude comes from the evaluation/assessment of curriculum materials.

The traditional emphasis of curriculum evaluation studies is on whether or not the educational materials or methods work. Sometimes the focus is on whether it works better than some other educational materials or methods. Evaluation (or assessment) studies are sometimes considered a separate issue from research. However, there are many similarities between research and evaluation studies if the exercise goes past a simple "does it work" and considers a "why does it work" or "not work" question. When designing educational materials or methods one question the educator should always ask is: why should you expect a difference? Without a theory this question cannot be answered. Curriculum development and instructional strategies should be based on theory and hence should be assessed by considering the theory as well as whether or not the materials work (e.g. see *12*).

The trouble with pure empiricism is that we don't have an idea about why what we do works or doesn't work. This hinders reproducibility of results. We reinvent successes and/or reproduce failures. An instructional tactic that an evaluation study has shown to be successful in one setting might not work in another. Without a theoretical base, the researcher has no strategy for developing explanations and possible modifications that will resolve the problem.

If otherwise proven instructional tactics or materials fail because they are used in combination with other incompatible practices, a theory-base will allow the identification of the nature of the incompatibility. It is difficult to implement a collection of random findings, which is what most practitioners are doomed to do. It's easier to implement research findings organized around a theory-base because it allows the analysis of both success and failure.

Learning Theory and Learning Preferences

Theories in science education are based on ideas about how people learn and what are the best conditions/ strategies/ materials to support that learning. But what if everybody learns differently? What are the ramifications? Is this an excuse to reject theory and ignore research findings?

There is a tension between the "students learn differently" position and the position that there are learning characteristics that are common to all learners. The traditional learning and developmental theories of researchers like Piaget imply this second position. There is an apparent contradiction between relativism – people are different and learn differently – and general learning theory – people learn the same way.

The "every learner is different" idea can be traced to the learning preferences literature. According to this literature, students have preferences in how they interact with instruction. Some prefer visual presentations, others verbal presentations; some prefer an inductive organization of information, others prefer a deductive organization; some prefer active interaction with materials, others prefer a more introspective interaction; and some prefer to cover content sequentially, while others like a holistic approach (*13*). Although these ideas would themselves seem to be theory-laden, they can be used to encourage an attitude of pure empiricism. The argument is to give the learner a preference test – there are many available – and teach accordingly. Unfortunately, this customized approach to instruction is not practical in a mass education setting. Furthermore, it may not encourage optimum learning. Felder (*13*) suggests that it would be better to offer different segments of instruction with different learning preferences in mind. Students need to practice different modes of learning, he argues, in order to develop into more powerful learners. Just because learners show a preference for one particular mode, doesn't mean they are incapable of learning by other modes and wouldn't profit by practicing them.

The tension between the "students learn differently" position and the position that there are learning characteristics that are common to all learners is a false one. Having a preference of learning strategies implies preference with how one interacts with content, educational materials, and people, not with how that material is assimilated, accommodated, and organized into the learner's mental structures. This process is an invariant in Piaget's theory (*14, 15*), for example, and is not necessarily in conflict with learning preference research.

Recent research in brain functioning indicates that learning is essentially based on brain physiology and chemistry (*16*). The mode of exposure doesn't change the brain's physiological or chemical response to the information being assimilated.

Unarticulated Theory

The previous sections of this chapter examined rationales for not basing research in chemical education on theory. However, even when not stated, or even consciously recognized by the researcher, theory is still implied by research decisions. Our educational practice is informed by our theories, i.e., our ideas about learning. A researcher who doesn't articulate an underlying theory about how learning takes place or what conditions are necessary for effective instruction is still operating and making theory-based educational decisions, i.e., decisions based on his or her ideas about learning. The problem with an unarticulated research theory is that it doesn't play an active role in the analysis and explanation of results, can cause a confusion of the critical with the trivial, and can lead to mistakes (*17, 18*).

Unarticulated theories of learning can be based on personal experience. These personal theories may be untested and of questionable validity. A researcher's personal experience is not necessarily a typical one. Basing research on an established theory has many advantages over using a personal theory. Research based on established theory provides coherence between research and practice and thus an opportunity to place education on firmer ground – to make it more effective. Using established theory as a basis for research makes research work more coherent and programmatic (*19*).

Roadblocks to Developing a Theory-base

Research can play a role in the practice of chemical education, but it has only to a limited extent. Why? Is it because the discipline of science/chemical education hasn't established a theory-base? At the present time classroom practice is informed by a combination of mentorship, anecdotal war stories, and history. What are the reasons for the research/practice gap? Why hasn't educational research influenced educational practice more?

There are several reasons why research in education is ignored. First of all, practitioners feel many research results are obvious. "Everybody knows that," is the argument. The research was unnecessary because common sense would have come to the same conclusion. This would imply that the practitioner was grounded in some theory-base that would lead one to think that the outcome was expected. This assumption, however, has been shown to be questionable. Wong (20) studied the perceived obviousness of the findings of research on teaching. Subjects had varying degrees of expertise in teaching (experienced teachers, teacher trainees, undergraduates in psychology, and undergraduates in engineering). Subjects were asked to rate the degree of obviousness (on a scale from extremely obvious to extremely unobvious) of paired sets of findings; one the actual research finding and the other its opposite. The findings chosen for the study were considered durable, replicable, and potent because they were derived from controlled experiments with large numbers of teachers and pupils from representative samples. Three treatments were studied. Subjects were asked to select the actual finding from each of the paired opposite possibilities and to rate the obviousness of that finding. In a second treatment subjects were asked to rate the obviousness of a single outcome that was either the actual or the direct opposite of the actual finding. The third treatment asked subjects to rate the obviousness of a single outcome when an explanation for the outcome was provided.

Wong found that consumers of research, no matter their level of expertise, are just as likely to identify the opposite finding as the actual finding as the correct result. Furthermore, they are likely to feel strongly that the findings (actual as well as opposite) are obvious. This lends support for the need to base educational practice on research rather than tradition. The study also indicates that explanations tend to increase the rated obviousness and therefore believability of research findings. This lends support to the idea that research should have a theory-base which can provide those explanations.

Research in education is also ignored because it is considered not relevant. The reputation is that researchers study problems that are too esoteric to be of interest to practitioners who have more practical needs. This is no longer true. We are now beginning to understand the process of education to a degree that allows us to make practical suggestions to practitioners. Support for this view is found in the collections of research findings that are now available (e.g. see 21). As one example, there is considerable evidence that inquiry-based laboratory activities have real educational advantages over more traditional approaches (22). And yet, as of 1997, less than 10% of college chemistry laboratories have adopted this approach (23). The same could be said for using cooperative groups (24-26) and concept mapping (27). All of these research-based practices are well grounded in theory and have great potential for improving instruction. All of these instructional strategies and tactics have had some influence on practice and have proved successful at improving teaching and learning. This alone is testimony to their relevancy. Because these instructional strategies and tactics are well grounded in research and are theory-based, they have the potential to change and improve chemical education practice in lasting ways.

Sources of Theory

Does chemical education have a theory-base or theory-bases? If so, what is it? An academic discipline is characterized as a field of study or a branch of knowledge that has a research base and theories that define its practice. A distinction is made between a pure scientific discipline like chemistry and an applied scientific discipline like medicine or engineering. Mature applied disciplines, like medicine, rely on theories derived from the pure sciences, from which they develop theories of their own. As an emerging applied discipline, chemical education has begun borrowing and modifying theories from more mature disciplines. Theories of how students learn chemistry concepts are related to other disciplines but are modified and adjusted to inform practice. What are the sources of these theories and what are the most promising theories to act as bases for research? Following is a discussion of the possible sources of theory that have been useful for research in chemical education.

Chemistry

Chemistry as a discipline has characteristics and concerns that are unique to it and that creates special educational problems. One important example is the emphasis on understanding chemical concepts at the macroscopic, sub-microscopic, and symbolic levels (*28, 29*). The visualization of atomic and molecular behavior is critical to the understanding of chemical systems. This area has a research history from other disciplines including the literature concerning three-dimensional perception and visualization (*30*). A more detailed discussion of molecular visualization can be found in chapter 6. (*31*)

Psychology

Learning and developmental theories from psychology are an obvious source of theory for science education research. Piaget's theory of intellectual development and his ideas concerning the assimilation, accommodation, and organization of knowledge has had a great influence on science education research for many years (*32, 33*).

Behaviorism has also been an influence on educational practice (*34*). The basic tenant of behaviorism is that human behavior can be studied scientifically without consideration of internal mental states. B. F. Skinner was a major figure in the behaviorist movement and was active in applying behaviorist principles to education (*35*). Skinner emphasized the use of conditioning and reinforcement to change behavior. Although Skinner argued against theory in psychological research (*36*), his approach has been used as a theory-base by educational

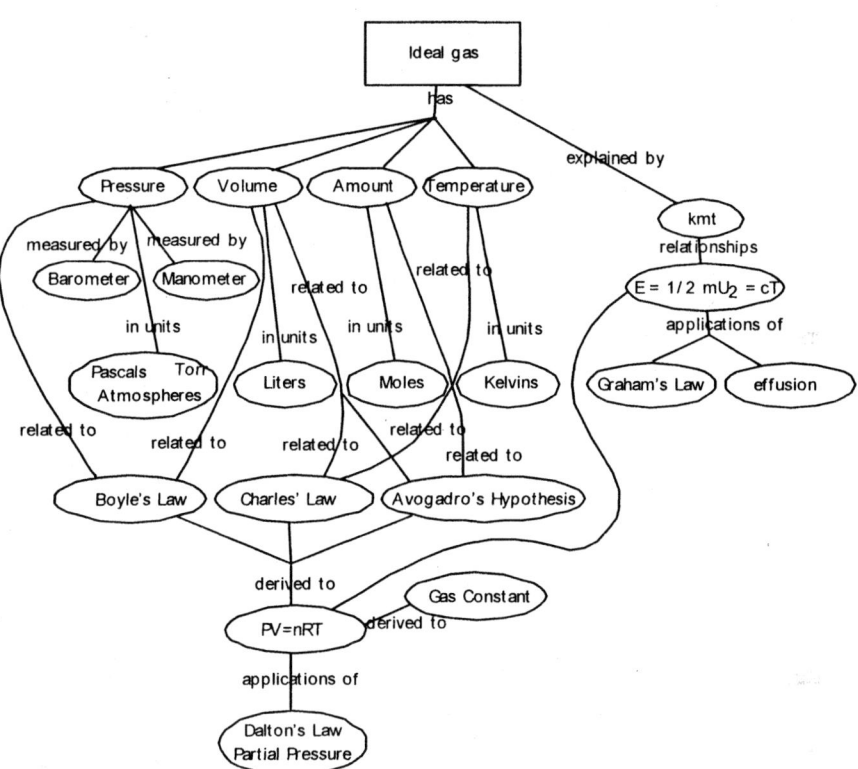

Figure 1. Sample concept map

assess what the students know or think they know about the subject being studied. These questions can be used to: pool data to be used to "invent" concepts in lecture, identify student misconceptions to be addressed in lecture, and review concepts needed as prerequisite knowledge for a lecture topic. The instructor can access student responses and use them to customize his/her lecture, to address specific students' misconceptions, to assess students' prerequisite knowledge, and to develop charts and graphs of student-generated observations that can be used to invent concepts (*58*). One of the goals of these questions is to encourage students to come to the lecture already thinking about the topic to be discussed. Many institutions use Just-in-Time Teaching in several different subject areas. Research results demonstrate improvements in retention rates and cognitive gains (*59, 60*).

Productive chemical/science education research programs that support beneficial instructional tactics have been based on both concept maps and JITT.

Related to Ausubel's theory, Gagné was similarly concerned with prerequisite knowledge and was also concerned with constructing hierarchical structures based on a task analysis (*37*). In Gagne's theory, one started with the final task, which was the desired outcome of instruction, and worked backwards by asking the question "what does the learner need to know to do this task." Eventually, one worked to the level where the learner operated. Instruction then started at that level and worked up to the final task.

Piaget's Theory of Intellectual Development (*61-63*)

Two aspects of Piaget theory have been of interest to science education researchers. The first of these has been called his stage model (*64*). Based on his observations, Piaget classified learners into categories of learning types based on their development. Learners at different developmental stages think in ways that are qualitatively (not quantitatively) different. Roughly determined by age, learners have intellectual abilities that progress from simple to more complex. The two highest stages, concrete and formal, are of most interest to researchers of school age learners. What is surprising is the observation that many college-age students and adults have not progressed to the highest stage of development and are only partially able to comprehend the formal (abstract) concepts that are the essence of college chemistry courses. Many research studies based on this observation have looked at why this is so and what can be done about it (*65-68*).

A second aspect of Piaget's theory that has been of interest to science education researchers has been called his functioning model. This model concerns how learners learn. According to Piaget, human beings have mental structures that interact with the environment. We assimilate or transform information from our environment into our existing mental structures. Our mental structures operate on the assimilated information and transform it in a process of accommodating to it. That is, we try to make sense of new information in terms of what we already know and try to fit the new information into existing mental structures (concepts). At the same time, what we know can be modified by the new information. Thus, information from the environment can transform our mental structures, while at the same time our mental structures can be transformed by the information. This change is driven and controlled by the process of disequilibration. Disequilibration is an incompatibility between the assimilated information and existing mental structures and requires either a change or accommodation of the mental structure or a change in the perception of the assimilated information. When our mental structures have accommodated the assimilated information, we are in a state of equilibrium and have reached an "accord of thought with things" (*14*). To use a chemical metaphor, disequilibration is the driving force pushing the system to equilibrium. In accommodating the new information, however, the altered mental structure can become disequilibrated with regards to related existing mental structures. The new structure must be organized with respect to the old structures in order to develop a new equilibrated organization. In other words, we must bring the "accord of thought with itself" (*14*).

Piaget's functioning model has been very fruitful as a theory-base for research. It has also been used as a rationale for inquiry-based instructional strategies (*22*). Inquiry-based instructional strategies are usually characterized as being question- and student-centered, and emphasizing hands-on activities used to provide data for the development of concepts. These strategies have in turn been the basis for further research. These strategies are usually thought of as being inductive in character. This can be seen as a contrast with instructional strategies, derived from Ausubel's theory, that are usually characterized as being deductive in nature. The contrast can be seen most clearly in the role taken by laboratory activities. In inquiry the role of the laboratory is to provide data from which concepts can be drawn. In the more traditional deductive instructional laboratories, the role is to verify a concept previously presented. One model of inquiry instruction originally proposed by Karplus and Thier (*69*) is called the learning cycle approach. The following analysis shows the influence of Piaget's functioning model on the learning cycle approach.

If learning spontaneously occurs through a process of assimilation, accommodation, and organization, then instruction could take advantage by arranging instructional activities to be compatible with this theory of learning. In order to facilitate assimilation, instructional activities should expose the learner to a segment of the environment that demonstrates the information to be accommodated. This should be followed by activities that help the learner to accommodate the information. Finally, in order to organize the accommodated information, activities should be developed that help the learner see the relationship between the new information and other previously learned information. The parallels between Piaget's functioning model, the Learning Cycle approach, and learning activities are illustrated in Table I (taken from *70*).

Table I. Piaget Functioning Model and the Learning Cycle

Piaget's Functioning Model	Learning Cycle Teaching Model	Learning Activities and Materials
Assimilation	Exploration	Data Collection & Analysis
Accommodation	Concept Invention	Conclusions and/or Interpretation
Organization	Application	Application Activities

SOURCE: Reproduced with permission from reference 70. Copyright 2005 Prentice Hall.

Another model of inquiry that is related to Piaget's functioning model and consistent with the learning cycle approach is that used in POGIL classrooms (71). The POGIL (process-oriented guided-inquiry learning) project has developed guided-inquiry instructional activities that use key scientific processes to develop scientific concepts. Students solve scientific problems in cooperative learning groups.

In addition to instructional materials and methods, other theory-bases have been derived from Piaget's theories. The one that has had the most influence in recent year has been constructivism (51). In its simplest terms, this theory states that the learner constructs concepts though a process of assimilation, accommodation, and organization similar to Piaget's ideas. There are many forms of constructivism that emphasize the intellectual or social aspect of learning (72).

One of the outcomes of the interest in Piaget's theory and the acceptance of constructivism is the observation that students come into our science classes with ideas about how the natural world works that are at odds with how experts think (73-75) . What is apparent is that these "misconceptions" are resistant to change by using normal instructional methods. Another observation is that these ideas are pervasive; that is, they are found to be common nonrandom ideas across cultures and populations. At first these ideas were studied by identifying them without developing any theory-base to try and explain them. The literature was filled with examples of student misconceptions for the "entertainment" of the reader. But this approach reached a dead-end and researchers finally started thinking about where these ideas came from and what to do about them. Research focused on: a) what students' conceptions are, b) why they arise, c) why are they resistant to change. At that point, the use of theory showed that misconception research could yield benefits for practice. One area where this has been of special benefit for chemical education has been the emphasis on helping students visualize the submicroscopic world (76).

Using a variety of theories, including that of Piaget, researchers proposed sources of the misconceptions including: alternate belief systems, historical misconceptions (a sort of an intellectual ontology recapitulates philology argument), reasoning ability, and instruction (75).

Conceptual change theory (77-79) was derived from constructivism and Piaget's functioning model (80). In order for learning to take place, the learner's misconceptions need to be confronted by the learner. The teacher needs to manufacture cognitive conflict (disequilibration) so that accommodation can take place. Conceptual change theory was also derived from the philosophical ideas of the nature of science proposed by Kuhn (50). According to Kuhn, what he called normal science continues unchallenged until its failure to explain a phenomenon. This anomaly leads to a scientific revolution; a change in the way the phenomena is explained. Conceptual change theory is parallel to these ideas.

Vygotsky's Social Constructivism (81, 82)

Vygotsky's contribution to ideas about learning is the Zone of Proximal Development (ZPD) (41), the gap between what a learner knows and what they are capable of knowing with instructional help. In order to bridge that gap, the learner needs to interact with a more capable individual. The role of that individual is to provide scaffolding or the intellectual support necessary for the learner to go beyond what he/she can accomplish on his/her own. Learning therefore takes place in a social environment. As a consequence, Vygotsky is thought of as a social constructivist and his theories have influenced social modes of instruction like cooperative learning, peer-led team learning, POGIL, and other group learning tactics.

The Problem of Sequence

To illustrate how a theory-base influences research and thus practice, consider the problem of how to decide the best sequence of curricular material in an instructional unit. One issue to consider is whether to use the nature of the subject matter or the nature of the learner as the central focus. Different learning theories might suggest different (but not necessarily incompatible) approaches.

Using an Ausubelian/ Gagnéan approach you, as an instructor, might draw a concept map or do a task analysis to help identify the topics covered and the order that seems to make sense from the point of view of the scientific discipline. Then you would sequence activities for the learners in the same order as the task analysis. The activities would start with a written exercise that would serve as an introduction to the instructional material that follows. At the end of the unit of study, students would be asked to construct a concept map to show the relationship among the ideas being studied. As a research project, you might test the effectiveness of your approach by comparing students who were exposed to this approach with students who received the same instruction, but with one of the elements of the approach removed. To test the effect of sequence, you could scramble the order of the activities and then test for student understanding.

From a Piagetian perspective, an instructional sequence might be arranged so that the more concrete – less abstract – ideas or approaches are presented first. By doing this you would be considering the intellectual reasoning ability of your students and adjusting the instruction to present it in a way to make it a more concrete experience. Since assimilation is a first step in learning, activities that present the learner with opportunities to assimilate would be used to introduce the topic. Thus, in designing a sequence of activities to teach a concept, a laboratory activity or simulation that would generate data related to the concept would come first, followed by a discussion designed to help the student accommodate the concept. During the discussion, if any misconceptions were generated, the learners would be asked to do activities that showed the inconsistency of these ideas. Finally, the student would be asked to do activities that relate the new concept to previously learned concepts. Such a strategy would take into account both the stage and functioning models of Piaget's theory. Research could be designed to test the success of this approach.

Using the approach of Vygotsky, a teacher would ascertain what students already know about the concept. Then they would design activities that support the students' progress toward the level of understanding that was the original goal. The activities could be done in a team or group setting with the teacher present to help the students. The sequence would be dictated by starting at the students current knowledge level of the topic and proceeding though a series of steps until the desired level of understanding was achieved. This process is called scaffolding. Systematic observation of the students' group interaction could help the teacher judge if the scaffolding provided was helping the students or if other scaffolding routes were needed.

The overall effect of the instructional materials could be assessed by taking into consideration the role played by the various components of instruction built into the materials. The theoretical framework used in the development of the materials would help the researcher identify the methods chosen to assess the materials. The theoretical framework would also influence the instruments and techniques used to collect data, and the variables chosen to judge the level of success of the materials. The criteria of what constitutes success itself might be determined by the choice of theory. The results of the assessment might also serve as a basis for judging the appropriateness of the theoretical framework itself.

These examples demonstrate how using a theory-base can be a powerful influence on practice through research. Having a theory-base helps the researcher identify critical variables to be manipulated and tested. Knowing these variables help the researcher define measurements to be taken, research methodologies to be utilized and analysis processes to be used. Results are more convincing if they are theory-based. Without theory, we just have a collection of interesting studies. With theory, these individual studies can be combined into an integrated whole that can address fundamental problems in learning and lead to targeted applications that will result in meaningful change.

References

1. DeBoer, G. E. *A History of Ideas in Science Education*; Teachers College Press: New York, 1991.
2. Yager, R. E.; Bybee, R.; Gallagher, J. J.; Renner, J. W. *Journal of Research in Science Teaching.* **1982**, *19*, 377-395.
3. Lincoln, Y. S.; Guba, E. G. *Naturalistic Inquiry*; Sage: Beverly Hills, CA, 1985.
4. Thomas, G., *British Educational Research Journal.* **2002**, *28*, 419-434.
5. Reichardt, C. S.; Cook, T. D. In *Qualitative and Quantitative Methods in Evaluation Research*; C. S. Reichardt and T. D Cook (Eds.); Sage, Beverly Hills, CA,: 1979.
6. Bretz, S. L. In *Nuts and Bolts of Chemical Education Research*; D. Bunce and R. Cole (Eds.); American Chemical Society Symposium Series, Washington, D.C.: 2007.
7. Sanger, M. J. In *Nuts and Bolts of Chemical Education Research*; D. Bunce and R. Cole (Eds.); American Chemical Society Symposium Series, Washington, D.C.: 2007.
8. Towns, M. H. In *Nuts and Bolts of Chemical Education Research*; D. Bunce and R. Cole (Eds.); American Chemical Society Symposium Series, Washington, D.C.: 2007.
9. Westmeyer, P. H., *Journal of Research in Science Teaching.* **1982**, *19*, 397-398.
10. Geelan, D. *From Undead Theories: Constructivism, Eclecticism, and Research in Education*; Sense Publisher: Rotterdam, the Netherlands, 2006.
11. Bennett, W. J. *What Works: Research About Teaching and Learning*; United States Department of Education: Washington, D. C, 1986.
12. Novak, J. D.; Musonda, D. *American Educational Research Journal.* **1991**, *28*, 117-153.
13. Felder, R. M. *Journal of College Science Teaching.* **1993**, 22, 286-290.
14. Piaget, J. *The Origins of Intelligence in Children*; Norton: New York, 1963.
15. Piaget, J. *Structuralism*; Harper and Row: New York, 1970.
16. Lawson, A. E., *Journal of Research in Science Teaching.* **1986**, *23*, 503-522.

17. Wynne-Edwards, V. C. *Animal Dispersion in Relation to Social Behaviour*; Oliver and Boyd: Edinburgh, 1962.
18. Wynne-Edwards, V. C. *Evolution Through Group Selection*; Blackwell: Oxford, England, 1986.
19. Novak, J. D. *A Theory of Education*; Cornell University Press: Ithaca, N.Y., 1977.
20. Wong, L. Y. *Journal of Educational Psychology.* **1995**, *87*, 504-511.
21. Gabel, D. L. *Handbook of Research on Science Teaching and Learning*; Macmillan: New York, 1994.
22. Lawson, A. E.; Abraham, M. R.; Renner, J. W. *A Theory of Instruction: Using the Learning Cycle to Teach Science Concepts and Thinking Skills;* Monograph, Number One; National Association for Research in Science Teaching: Kansas State University, Manhattan, KS, 1989.
23. Abraham, M. R.; Cracolice, M. S.; Graves, A. P.; Aldahmash, A. H.; Kihega, J. G.; Palma Gil, J. G.; Varghese, V., *Journal of Chemical Education* (& *JCE Online*, URL http://jchemed.chem.wisc.edu/). **1997**, *74*, 591-594.
24. Johnson, D. W. *Active Learning: Cooperation in the College Classroom*; Burgess: Minneapolis, MN, 1991.
25. Johnson, D. W.; Johnson, R. T.; Holubec, E. J. *Cooperation in the Classroom*; Burgess: Minneapolis, MN, 1993.
26. Johnson, R. T., *Journal of Research in Science Teaching.* **1976,** *13*, 55-63.
27. *Journal of Research in Science Teaching vol 27, issue 10;* Novak, J.; Wandersee, J. E. Eds.; 1990.
28. Gabel, D. L.; Samuel, K. V.; Hunn, D. *Journal of Chemical Education.* **1987**, *64*, 695-697.
29. Johnstone, A. H. *Chemistry In Britain.* **1982**, *18*, 409-410.
30. Paivio, A. *Mental Representations: A Dual Coding Approach*; Oxford University Press: New York, 1986.
31. Williamson, V. M. *In Nuts and Bolts of Chemical Education Research;* D. Bunce and R. Cole (Eds.); American Chemical Society Symposium Series, Washington, D.C.: 2007.
32. Good, R.; Mellon, E. K.; Kromhout, R. A. *Journal of Chemical Education.* **1978**, *55*, 688-693.
33. Herron, J. D. *Journal of Chemical Education.* **1975**, *52*, 146-150.
34. Skinner, B. F. *Harvard Educational Review.* **1954**, *24*, 86-97.
35. Skinner, B. F. *Walden Two*; Macmillan: New York, 1948.
36. Skinner, B. F. *Psychological Review.* **1950**, *57*, 193-216.
37. Gagné, R. M. *The Conditions of Learning and Theory of Instruction*, 4th ed.; Holt: New York, 1985.
38. Markle, S. *Good Frames and Bad.* 2nd ed.; Wiley: New York, 1969.
39. Gagne, R. M.; Briggs, L. J.; Wager, W. W. *Principles of Instructional Design*; Harcourt Brace Jovanovich: Fort Worth, 1992.

40. Ausubel, D. P.; Novak, J. D.; Hanesian, H. *Educational Psychology: A Cognitive View*. 2nd ed.; Holt, Rinehart and Winston: New York, 1968.
41. Vygotsky, L. S. *Thought and language*; M.I.T. Press: Cambridge, Massachusetts, 1962.
42. Brown, R. *Group Processes. Dynamics within and between groups*; Blackwell: Oxford, 1988.
43. Aikenhead, G. S., *Studies in Science Education.* **1996**, *27*, 1-52..
44. Aikenhead, G. S., *Science Education.* **1997**, *81*, 217-238.
45. Brickhouse, N. W. *Journal of Research in Science Teaching.* **2001**, *38*, 282-295.
46. Stanley, W. B.; Brickhouse, N. W. *Science Education.* **2001**, *85*, 35-49.
47. Cobern, W. W. *Journal of Science Teacher Education.* **1991**, *2*, 45-47.
48. Martin, M. *Concepts of Science Education: A Philosophical Analysis*; Scott, Foresman: Glenview, IL, 1972.
49. Lederman, N. G. *Journal of Research in Science Teaching.* **1992**, *29*, 331-359.
50. Kuhn, T. S. *The Structure of Scientific Revolutions*, 2nd ed.; University of Chicago Press: Chicago, 1970.
51. Bodner, G. M. *Journal of Chemical Education.* **1986**, *63*, 873-878.
52. Ausubel, D. P. *Journal of Educational Psychology.* **1960**, *51*, 267-272.
53. Ausubel, D. P. *The Psychology of Meaningful Verbal Learning*; Grune & Stratton: New York, 1963.
54. Novak, J. D.; Gowin, D. B., *Learning How to Learn*; Cambridge University Press: Cambridge, England, 1984.
55. Francisco, J. S.; Nakhleh, M. B.; Nurrenbern, S. C.; Miller, M. L. *Journal of Chemical Education.* **2002**, *79*, 248.
56. Markow, P. G.; Lonning, R. A. *Journal of Research in Science Teaching.* **1998**, *35*, 1015-1029.
57. Nicoll, G.; Francisco, J. S.; Nakhleh, M. B. *Journal of Chemical Education.* **2001**, *78*, 1111.
58. Novak, G. M.; Patterson, E. T.; Gavrin, A. D.; Christian, W., *Just-In-Time Teaching: Blending Active Learning With Web Technology*; Prentice-Hall: Upper Saddle River, NJ, 1999.
59. Linneman, S.; Plake, T. *Journal of Geoscience Education.* **2006**, *54*, 18-24.
60. Marrs, K. A.; Blake, R. E.; Gavrin, D. *Journal of College Science Teaching.* **2003**, *33*, 42-47.
61. Brainerd, C. J. *Piaget's Theory of Intelligence*; Prentice Hall: New Jersey, 1978.
62. Evans, R. *Jean Piaget: The Man and His Ideas*; E. P. Dutton: New York, 1973.
63. Renner, J.; Stafford, D.; Lawson, A.; McKinnon, J.; Friot, E.; Kellogg, D. *Research, Teaching, and Learning With the Piaget model*; University of Oklahoma Press: Norman, OK, 1976.

64. Inhelder, B.; Piaget, J. *The Growth of Logical Thinking: From Childhood to Adolescence*. Basic Books: New York, 1958.
65. Cantu, L. L.; Herron, J. D. *Journal of Research In Science Teaching*. **1978**, *15*, 135-143.
66. Herron, J. D. *Journal of Chemical Education*. **1975**, *52*, 146-150.
67. Herron, J. D. *Journal Of Chemical Education*. **1978**, *55*, 165-170.
68. Ward, C. R.; Herron, J. D. *Journal of Research in Science Teaching*. **1980**, *17*, 387-400.
69. Karplus, R.; Thier, H. *A New Look at Elementary School Science*; Rand-McNally: Chicago, 1967.
70. Abraham, M. R. In *Chemists' Guide to Effective Teaching*; N. J. Pienta, M. M. Cooper, and T. J. Greenbowe, (Eds); Prentice Hall, Upper Saddle River, N. J.: 2005.
71. Farrell, J. J.; Moog, R. S.; Spencer, J. N. *Journal of Chemical Education*. **1999**, *76*, 570-574.
72. Geelan, D. R. *Science & Education*. **1997**, *6*, 15-28.
73. Abraham, M. R.; Grzybowski, E. B.; Renner, J. W.; Marek, E. A. *Journal of Research in Science Teaching*. **1992**, *29*, 105-120.
74. Abraham, M. R.; Williamson, V. M.; Westbrook, S. L. *Journal of Research In Science Teaching*. **1994**, *31*, 147-165.
75. Haidar, A. H.; Abraham, M. R. *Journal of Research in Science Teaching*. **1991**, *28*, 919-938.
76. Tasker, R.; Dalton, R. *Chemistry Education Research and Practice*. **2006**, *7*, 141 - 159.
77. Driver, R. *European Journal of Science Education*. **1981**, *3*, 93-101.
78. Driver, R.; Easley, J., *Studies in Science Education*. **1978**, *5*, 61-84.
79. Hewson, P. W. *European Journal of Science Education*. **1981**, *3*, 383-396.
80. Posner, G. J.; Strike, K. A.; Hewson, P. W.; Gertzog, W. A., *Science Education*. **1982**, *66*, 211-227.

Chapter 6

The Particulate Nature of Matter: An Example of How Theory-Based Research Can Impact the Field

Vickie M. Williamson

Texas A&M University, College Station, TX 77843-3255

The Particulate Nature of Matter is vital to understanding chemistry. Chemists explain phemonena in terms of particle behavior. Several chemical education research studies have helped expand the theory of how students learn about particle behavior. Early studies established the lack of student understanding of particle action, while later studies examined treatments or interventions to help students think in terms of particles. These later studies led to a number of implications for the chemistry classroom and our understanding of how students build mental models to visualize particle behavior in chemical and physical phemonena.

Introduction

Theory-based research can have wide impacts. This chapter will describe the theories underlying the Particulate Nature of Matter (PNM). Examples of descriptive and intervention studies involving the PNM will be given, along with their impact on the field of chemical education research, classroom practice, standarized examinations, funding agencies, and conferences. The chapter serves as an example of how theory-based research can cause dramatic changes in instructional practices and have a significant impact on learning theory.

Theoretical Framework of the Particulate Nature of Matter

What are the different views of matter?

A chemist's work, like that of any scientist, involves observation and experimentation. The view of matter and its changes that can be seen in the laboratory is considered the macroscopic component of chemistry (*1*). Chemists use symbols and mathematics to describe their observations and findings (the symbolic component). Finally, chemists relate what they have observed to the action of particles (atoms and molecules) in the submicroscopic or particulate component. Johnstone (*1*) explained that chemists consider these multiple components or representations simultaneously and can easily move between the representations as needed. The Particulate Nature of Matter (PNM) is the underlying theory for the particulate component of chemistry that Johnstone described, the idea that matter is composed of particles. For more information concerning the PNM, see an earlier work by Johnstone (*2*) or Gabel's chapter in the *Chemist's Guide to Effective Teaching* (*3*).

Mental Models of Novices and Experts

Mental models are the pictures that we 'see' in our mind when we are thinking about something. The terms mental pictures, mental models, or visualizations all refer to these pictures in the mind. A person can have a mental model of a concrete, macroscopic thing that they have have seen in the past (e.g., a beaker), but the more difficult mental models are those of an abstract thing that the person has not seen and cannot see (e.g., molecules). The terms concrete and abstract or formal come from the learning theory of Jean Piaget (*4-8*)..It was Piaget's finding that at a certain developmental level, students required experience with concrete objects to gain understanding and that they could think about that experience via mental pictures of the concrete objects. At a later level, students could learn through experience with abstract thought or ideas, with mental models forming much more quickly. Student could think about their thinking.

Constructivism builds on the work of Piaget and includes the ideas from von Glasersfeld (*9*) that our knowledge is constructed to fit ultimate reality rather than to match it. Constructivism is the belief that: (a) knowledge is constructed from interactions with people and materials, not transmitted, (b) prior knowledge impacts learning, (c) learning, especially initial understanding, is context specific, and (d) purposeful learning activities are required to facilitate the construction or modification of knowledge structures (*10-11*). Thus, learning is an active process in which the individual builds or constructs meaning from experiences and events which must be integrated into existing conceptual frameworks.

With learning theory, the type of mental models that are held are not the same for all persons. For Johnson-Laird (*12*), our knowledge depends on our abilty to construct mental models from our conceptual frameworks, with which we can use to reason. Larkin (*13*) described the differences between the mental models of experts and novices. The type of visualization or mental model constructed by experts differs from those of novices. Novices usually have incomplete or inaccurate models, while those built by experts include both sensory, macroscopic data from the physical world and formal abstract dimensions of the phenomena.

Further research was conducted to describe the differences between experts and novices. Kozma & Russell (*14*) compared sorting and transformation tasks between chemistry novices and experts. They provided novices and experts with a range of representations, including video segments (macroscopic), graphs (symbolic), animations (particulate), and equations (symbolic). The researchers found that novices made smaller groupings and more often grouped items with the same type of representation. The novices gave explanations for their groupings based on surface features. Experts formed larger groups with multiple types of representations and gave reasons for the grouping that were more conceptual. The authors related these differences to the experts' more complete mental models.

Johnson (*15*) described four distinct types of particulate mental models that move along the novice–expert continuum. Students using the first type of mental model have no idea of particles. They see matter as continuous. In the second type, students draw particles, but see the particles as something separate from the substance. For example, they draw molecules inside the sugar cube or draw water molecules, but will say that water is in between the drawn molecules. In the third type of mental model, students believe that the particles make up the substance, but attribute the macroscopic properties of the substance (the element or compound) to the individual particles. For example, they draw water molecules in steam as wavy or sodium atoms as silver. In the fourth type, students understand that the particles make up the substance AND that the macroscopic properties are attributed to the collection of particles, not to individual particles. As instructors, we want to move our students towards more complete expert mental models.

Research Studies Concerning the Particulate Nature of Matter

Misconception/alternate conception studies

A large number of studies established that students hold misconceptions concerning the PNM. Although studies were completed with almost every topic in beginning chemistry, only a few representative studies will be discussed to illustrate the progression of the research in this area. Early instructional practice was to teach chemistry with an emphasis on algorithmic problem solving, believing that students would understand the PNM if they could solve mathematical problems.

Novick & Nussbam (*16*) conducted a cross-age study in early work on the topic to show that students hold misconceptions about nature and behavior of particles. They found that many misconceptions are not overcome with age. For example, over 60% of high school and university students did not picture empty space between gas particles, and more than 50% of these students did not show uniform distribution of gas particles in a closed flask. In addition, less than 50% of these students correctly indicated that the uniform particle distribution was due to constant particle motion. The study recommended that instructors should be aware of such misconceptions by soliciting students' ideas concerning chemistry concepts and should use this knowledge to prepare curriculum materials. At this point, it was not known exactly what one could use to combat these misconceptions dealing with phenomena that could not be seen.

In another key study, Ben-Zvi, Eylon, & Silberstein (*17*) investigated the understanding of 300 high school chemistry students and found that 66.3% held continuous, rather than particulate, views of matter. In their study, students incorrectly attributed the properties of the element (e.g., color, malleability, compressibility, expansion on heating, odor, etc.) to an individual atom of the element.

A series of studies were conducted in the late 1980's and early 1990's which focused on algorithmic problem solving and conceptual understanding. Conceptual understanding involved the PNM in drawings, explanations of particle behavior, or predictions of what the system would do under other conditions. These studies resulted in data that illustrated the lack of understanding of the PNM, even for students who could solve algorithmic problems and lead to a call for BOTH conceptual and algorithmic instruction (*18-21*). Previously, the symbolic level was the domain of the lecture course, while the macroscopic level was the domain of the laboratory course. The idea was now born that students should be instructed at the particulate level, in addition to the symbolic and the macroscopic levels.

Researchers began to propose attitributes for the nature of particulate instruction. Gabel, Samuel, & Hunn (*22*) called for an increased emphasis on the PNM in introductory courses, the need for careful depiction of particles used in instruction, and the need to depict physical phenomena in terms of the PNM. The presevice teachers in their study were given pictures of atoms and molecules represented by circles of various sizes and shades, and then were asked to draw a new picture of the resulting physical or chemical change. Subjects ignored conservation of particles and spacing of particles to illustrate the phase of matter in over 50% of the cases. Analysis showed that their formal reasoning ability was related to their PNM understanding and that their visual rotation ability or number of completed mathematics courses had no effect on their PNM understanding.

Haider & Abraham (*23*) analyzed of the nature of students' alternative conceptions and their use of the particulate theory when questions were worded in scientific language vs. everyday language. The researchers found a significant difference between students' applied and theoretical knowledge in answering these questions. For example, students responded differently when asked about a teaspoonful of sugar being stirred in water, than they did when asked about the mixing of sugar and water molecules. The authors proposed that students compartmentalize their knowledge and only use particulate terms when prompted to do so by the curriculum materials or teacher. The researchers warned that instructors must take care in the language they use in instruction to elicit the desired response from students.

In a later cross-age study, Abraham, Williamson, Westbrook (*24*) investigated the conceptions that students held at the end of middle school physical science, high school chemistry, and college-level general chemistry. The authors examined five concepts that were taught at each level (chemical change, dissolution of a solid in water, conservation of atoms, periodicity and phase change). The authors found that both reasoning ability and experience with the concepts (grade level) accounted for the understanding of the concepts tested. However, misconceptions concerning the PNM in these processes were present at all levels. For example, 9% of the 100 college, 17% of the 100 high school, and 9% of the 100 junior high students (11.7% of the total sample) believed that sugar particles sank to the bottom or floated instead of evenly mixing when sugar is dissolved in water. Further, students tended to resist using atomic and molecular explanations. The voluntary use of atoms and molecules increased with age for this same item, which did not use these terms; 13% of the junior high, 30% of the high school, and 46% of the college students (29.7% of the total sample) used atoms and molecules in their explanations. The use was not always correct; many students referred to the "sugar atoms". The authors recommended that more effective teaching methods should be developed to help students link experience-based (macroscopic) observations with the atomic and molecular (particulate) models used by chemists to explain the phenomena. Recommendations included putting more emphasis on how concepts are developed and modified so that students will feel comfortable with changing their own concepts in the face of evidence and that more emphasis should be on concept-based, rather than fact-based curriculm.

Understanding the PMN has been a persistent area of difficulty for students. Nahkleh and Samarapungavan (*25*) in their investigation of the understandings of elementary students used in-depth interviews to probe student understanding of the phases of matter, phase changes, and dissolving. Most elementary students had ideas about matter as particles, but the particles had the macroscopic characteristics of the material. A small number of the elementary student (20%

or 3 of the 15) held ideas that matter was continuous, while the same number held PNM views. The explanations given for phase changes and dissolving were consistent with the elementary students ideas about matter. In a similar study,

Nakhleh, Samarapungavan, and Saglam (*26*) investigated the understanding of middle school students. Most middle school students (67% or 6 of the 9) showed some understanding of the PNM and knew that matter was composed of atoms and molecules; however this understanding was not consistently applied to the different examples used in the interview. Further, PNM ideas were not used to explain properties of matter, phase change, or dissolving. The authors proposed middle school students were in a state of transition from a macroscopic to a particulate view of matter and that the fragmentation and localization of understanding was due to the difficulty that exists in this transition. Suggestions for instruction included first investigating the substances that students can identify as particulate in nature (e.g., water or helium), then moving to granular substances (e.g., sugar or salt), and finally to nongranular solids (e.g., metals or wood).

In summary it has been demonstrated through research studies that chemistry students hold many misconceptions and have little understanding of a wide variety of concepts concerning the PMN (*16, 18-22, 24-28*). A large number of these difficulties are caused by the students' application of macroscopic explanations from their everday experience to the particulate concepts (*23*) or by the students' inability to visualizae, diagram, or depict the behavior of particles (*17, 22*).

Treatment/intervention studies

Following on the heels of the previous studies which called for changes in teaching strategies, experimental research was conducted to test the effectiveness of an intervention or treatment. A number of interventions have been shown to help students understand the PNM. In gaining this understanding, students have been shown to develop more expert-like mental models of the chemical phenomena and our understanding of good instructional practices have changed. Gabel and Sherwood (*29*) found that students who manipulated physical models of particulate level interactions performed significantly better on solving general chemistry problems than students who only saw their instructors demonstrating the models. Gabel, Hitt, and Yang (*30*) found increases in PNM understanding using Play-Doh models, which prospective teachers manipulated.

Ben-Zvi, Eyon, & Silberstein (*17*) in the second part of their paper, developed a new teaching strategy to develop the atomic model that was tested with 540 high school chemistry students. This new strategy took a historical approach in presenting the atom as a changing model when new facts are discovered. The authors found that after instruction, 43.7% of the experimental group had an acceptable understanding of the nature of the atom and the structure of matter, compared to only 18.4% of the control group. They found that 80-90% of the students with correct ideas about the character of an atom also correctly visualized the structure of compounds. At the same time students with little understanding of the character of an atom were evenly divided between correct and incorrect visualizations of compounds. The authors argued that these results suggest that it is worthwhile to help students internalize the correct atomic view.

A number of studies have proposed benefits from the use of animations on a wide variety of chemistry topics (*31, 32*). For example, Williamson & Abraham (*33*) found that college chemistry students who were exposed to short PNM animations during lecture performed significantly better on conceptual, particulate questions than did the control group who viewed only static visuals of particles. This study involved solids, liquids, gases, and solution reactions. The authors suggested that the lack of understanding of chemistry concepts may be linked to the students' inability to build complete mental models that visualize particulate behavior. This study proposed that the dynamic nature of computer animations enabled deeper encoding and more expert-like mental models, as compared to those developed with static visuals. Also see Sanger (*34*) for a more complete discussion of animations.

There is some evidence that a gender gap may exist concerning PNM understanding. Yezierski and Birk (*35*) found a gender gap on their PNM pre-test. However, after treatment with animations, not only did the treatment group perform significantly better on the PNM assessment, but the gender gap had disappeared. The researchers called for frequent use of particle-level animations, along with accompanying discussions concerning the animations and the interpretations of what students have observed. Further suggestions include training teachers to determine student misconceptions and to design interventions using accessible PNM animations.

Animations have been used with other visualizations. Sanger & Badger (*36*) found that electron density plots of simple molecules aided student understanding of polarity and miscibility at the particle-level when used with animations. Russell, et. al. (*37*) found that the use of simultaneous-synchronized macroscopic, particulate, and symbolic representations enhanced the teaching and learning of chemical concepts.

Allowing students to draw, build, choose, view, and rotate models of molecules have been shown to help students' understanding. Harrison & Treagust (*38*) used student drawings to evaluate student understanding of atoms. They proposed that instructors should use multiple models or representations in instruction and that the visualization techniques selected should be appropriate to the cognitive ability of their students. Instructors should gradually challenge students to use more abstract models, with drawing as a way for the instructor to access the mental models held by students. Wu, Krajcik, Soloway (*39*) investigated the use of a computer-based visualization tool, eChem, that allowed students to build molecular models and view multiple images at one time (2D and 3D). The researchers concluded that computerized models may serve as a vehicle for students to generate mental images.

There is some evidence that asking students to storyboard or to make their own animations helps in the formation of particulate mental models. In this way students are illustrating their own mental models. Milne (*40*) described an activity using only note cards and pencils to create an 'animation' using individual drawings that are assembled so to provide a dynamic chemical reaction as seen through a flipbook. Schank & Kozma (*41*) found that when students used *Chemsense,* a molecular drawing and animation tool, the students were significantly better at representing chemical phenomena at the particulate level. The Chemsense tool allows students to create their own animations. Additionally, they found that students were more focused on the dynamic nature of chemical reactions.

Velazquez-Marcano et. al. (*42*) found in their study of college general chemistry that both a particulate animation and a macroscopic demonstration of the phenomena were needed for the maximum effect when students were asked to predict the outcome of fluid experiments at the macroscopic scale. Both the particulate and the macroscopic treatments were needed; however, the order of the visualizations did not matter for significantly better scores. There was no gender effect found in this study. The authors called for the use of multiple types of visualizations in instruction.

In summary, a number of representative studies have suggested that students' understanding of the PNM can be enhanced by using physical models (*29, 30*), student drawings (*38*), computer programs that generate molecules which can be rotated and represented in various ways (*38, 39*), animations (*33, 36*), and student drawings or animations (*40, 41*). Further, there is consensus in the literature that more than one visualization technique should be used to help students create mental images of the PNM (*37, 38, 42*). For more information, see the summary of research findings on visualization in chemistry by Wu & Shah (*43*).

Influence of Theory-Based Research

The theory-based research studies have changed our understanding of good teaching practices. Before the problems with student understanding of the PNM were discovered, teaching was done at only the symbolic or algorithmic levels, with some macroscopic teaching occurring in the laboratory or through classroom demonstrations. The implications from the misconception and intervention studies is that understanding the Particulate Nature of Matter depends on the mental models held by the individual. These mental models can be developed by using physical models, drawings, computer programs, animations, and student-generated drawings/animations. The materials must be used in a number of settings and more than one visualization technique should be used. Dynamic models that can be presented in an animation seem to have an added value over the static images, as used in transparencies, etc. Additionally, these efforts to use multiple particulate visualizations must be coupled with other representations at the macroscopic and symbolic/mathematical levels.

The influence of the research findings dealing with the PNM has extended to a number of areas. Understanding at the particule level is acknowledged as vital for a deep understanding of chemistry by standardized examinations, research conferences, and classroom practice. The Examinations Institute of the American Chemical Society's Division of Chemical has offered conceptual examinations for the two-semester sequence of general chemistry since 1996. Many of the conceptual questions deal with particle behavior through drawings or words. Additionally, the regular examinations for general chemistry have included increasingly more particulate questions. For example, about 40% of the questions from the ACS Exams Institute's 2001 Conceptual Examination are particulate questions, while the 2002 First-Term General Chemistry Exam contains about 13% particulate questions.

Gordon Research Conferences have hosted a conference on Visualization and Science Education since 2001. These visualization conferences have focused on visualization techniques to be used in the classroom to help students form more expert-like mental models. In chemistry, this means the models of particle behavior in chemical processes and reactions. The National Science Foundation has promoted visualization in the classroom by funding a number of workshops on the topic. José & Williamson (*44*) report one such workshop involving chemists, chemical educators, and software developers who worked through complex ideas such as what is the role of particulate animations in the classroom and what are the characteristics of a good animation.

The educational practices today are very different from those first described. Changes began slowly as the evidence from the research studies began to accumulate. These include the recognition that teaching must include conceptual methods which promote PNM understanding. Classroom interventions to promote PNM understanding include the use of physical models (model kits, magnets, marshmallows, etc.), role playing molecules, and computer models, including those that simply rotate, animations of processes, student-generated drawings or animations, and interactive computer models in which students control variables (*45*). Most textbook ancillaries now include clips of animations depicting particle behavior as well as incorporating particulate drawings in the textbooks. Many of these also include multiple representations (macroscopic, particulate, symbolic). Electronic homework systems and exam banks also use particulate questions. Animations of particle behavior are available on the Internet (for one example, see http://www.chem.iastate.edu/group/Greenbowe/ sections/projectfolder/animationsindex.htm).

Summary

This chapter serves as an example of how research studies that are grounded in a theory base can expand the extent to which a theory is used or accepted. The studies cited here are but a few of those dealing with the PNM and were meant only as examples. Good research builds on the work of others to develop a weight of evidence on a topic. The difficulties that students have understanding the Particulate Nature of Matter is well documented. Further, much is known about the nature and cause of these difficulties, along with a number of teaching strategies that will help students overcome the difficulties. The knowledge gained in the area has influenced policy-makers, curriculum developers, animators, and instructors. Having a theoretical basis for research can more completely and quickly influence the field of teaching and learning. This body of research serves as an example of how research which has a theoretical underpinning has resulted in dramatic changes to the educational theory regarding how students develop PNM mental models. It is essential for the advancement of the field to not only to investigate how and when an educational practice works, but to also explain its action by a robust theory of learning or to use the findings to further define the learning theory.

References

1. Johnstone, A.J. *J. Chem. Ed.* **1993**, *70*, 701-705.
2. Johnstone, A.J. *J. Comp. Assist. Learn.* **1991**, *17*, 701-703.
3. Gabel, D. Enhancing students' conceptual understanding of chemistry through integrating the macroscopic, particle, and symbolic representations of matter. In *Chemist's Guide to Effective Teaching*; Pienta, N. J.; Cooper, M. M.; & Greenbowe, T. J., Eds. ; Pearson: Upper Saddle River, NJ; 2005; pp 77-88.
4. Herron, J. D. *J. Chem. Ed.* **1975**, *52*, 146-150.
5. Herron, J. D. *J. Chem. Ed.* **1978**, *55*, 165-170.
6. Nurrenbern, S. C. *J. Chem. Ed.* **2001**, *78*, 1107-1110
7. Piaget, J. ; Inhelder, B. *The Psychology of the Child.* Basic Books: New York City, NY 1969.
8. Piaget, J. *The Development of Thought: Equilibrium of Cognitve Structures.* Viking: NY 1977.

9. von Glasersfeld, E. *Radical Constructivism: A way of Knowing and Learning*. Falmer: Washington, DC. 1995.
10. Osborne, R.J.;Wittrock, M.C. *Sci. Educ.* **1983**, *67*, 489-508.
11. Bodner, G.M. *J. Chem. Ed.* **1986**, *63*, 873-878.
12. Johnson-Laird, P.N. Mental Models. In *Foundations of Cognitive Science;* Posner, M.I., Ed.; MIT Press: Cambridge, MA; 1989, 469-499.
13. Larkin, J.H. The role of problem representation in physics. In *Mental Models*; Genter, D. & Stevens, A.; Eds.; Lawrence Erlbaum Associates: Hillsdale, NJ; 1983; pp. 75-98
14. Kozma, R.B.; Russell, J. *J. Res. Sci. Teach.* **1997**, *34*, 949-968
15. Johnson, P. *Int. J. Sci. Educ.*, **1998**, *20*, 393-412.
16. Novick, S.; Nussbaum, J. *Sci. Educ.* **1981**, *65*, 187-196.
17. Ben-Zvi, R.; Eylon, B.R.; & Silberstein, J. *J. Chem. Educ.* **1986**, *63*, 64-66.
18. Nurrenbern, S. C., & Pickering, M. *J. Chem. Ed.* **1987**, *64*, 508-510.
19. Pickering, M. *J. Chem. Ed.* **1990**, *67*, 254-255.
20. Sawrey, B. A. *J. Chem. Ed.* **1990**, *67*, 253-254.
21. Nakhleh, M. B. *J. Chem. Ed.* **1993**, *70*, 52-55.
22. Gabel, D. L.; Samuel, K. V.; Hunn, D. *J. Chem. Educ.* **1987**, *64*, 695-697.
23. Haidar, A.H. & Abraham, M.R. *J. Res. Sci. Teach.* 1991, *28*, 919-938.
24. Abraham, M. R.; Williamson, V. M.; & Westbrook, S. L. *J. Res. Sci. Teach.* **1994**, *31*, 147-165
25. Nahkleh, M.B.; Samarapungavan, J. . *J. Res, Sci. Teach.*, *36 (7)*, **1999**, 777-805.
26. Nahkleh, M.B.; Samarapungavan, J.; Saglam, Y. . *Res, Sci. Teach., 42 (5)*, **2005**, 581-612.
27. Andersson, B. *Sci. Educ.* **1986,** *70*, 549-563.
28. Osborne, R. J., & Cosgrove, M. M. . *J. Res. Sci. Teach.* **1983**, *20*, 825-838.
29. Gabel, D.; Sherwood, R. *J. Res. Sci. Teach.* **1980**, *17*, 75-81.
30. Gabel, D. L., Hitt, A. M., & Yang, L. *Changing prospective elementary teachers' understanding of the macroscopic, particulate, and symbolic representations of matter using Play-Doh models.* Presented at the annual meeting of the National Association for Research in Science Teaching, Philadelphia, PA, 2003.
31. Kelly, R. M., Phelps, A. J., & Sanger, M. J. *The Chem. Educ.* **2004**, *9*, 184-189.
32. Sanger, M. *J. Chem. Ed.* **2000**, *77*, 762-766.
33. Williamson, V.M.; Abraham, M.R. *J. Res. Sci. Teach.*, **1995**, *32*, 521-534
34. Sanger, M. J. Computer animations of chemical processes at the molecular level. In *Chemist's Guide to Effective Teaching, Volume 2*; Pienta, N. J.; Cooper, M. M.; & Greenbowe, T. J., Eds. ; Pearson: Upper Saddle River, NJ; in press.
35. Yezierski, E. J., & Birk, J. P. *J. Chem. Educ.* **2006**, *83*, 954-960.
36. Sanger, M.J. & Badger, S.M. *J. Chem. Ed.* **2001**, *78*, 1412-1416.

37. Russell, J.W.; Kozma, R.B.; Jones, T.; Wykoff, J.; Marx, N.; Davis, J. *J. Chem. Ed.*, **1997**, *74*, 330-334
38. Harrison, A.G.; Treagust, D.F. *School Sci. and Math.* **1998**, *98*, 420-429.
39. Wu, H.; Krajcik, J.S.; Soloway, E. *J. Res. Sci. Teach.*, **2001**, *38*, 821-842.
40. Milne, R.W. *J. Chem. Ed.* **1999**, *76*, 50-51.
41. Schank, P.; Kozma, R. *J. Computers in Math. and Sci. Teach.* **2002**, *21*, 253-279.
42. Velazquez-Marcano, A; Williamson, V. M.; Ashkenazi, G.; Tasker, R.; Williamson, K. C. *J. of Sci. Educ. and Tech.* **2004**, *13*, 315-323.
43. Wu, H., Shah, P. *Sci. Educ.* **2004**, 88(3), 465-492.
44. José, T.J.; Williamson, V.M. *J. Chem. Ed.* **2005**, *82*, 937-943.
45. Williamson, V.M., & Jose, T.J. Using Visualization Techniques in Chemistry Teaching. In *Chemist's Guide to Effective Teaching, Volume 2*; Pienta, N. J.; Cooper, M. M.; & Greenbowe, T. J., Eds. Pearson: Upper Saddle River, N.J.; in press.

Chapter 7

Qualitative Research Designs in Chemistry Education Research

Stacey Lowery Bretz

Department of Chemistry and Biochemistry, Miami University, Oxford, OH 45056

Qualitative research methodologies are uniquely suited to exploring the mechanisms of teaching and learning chemistry. This chapter examines the issue of the fit between research question and research design, as well as provides an overview of the procedures used to collect data in qualitative research. Different traditions within qualitative research and examples of research studies within these traditions are discussed. The chapter concludes by examining the realities of conducting qualitative research in chemistry education given the traditional methods of research in chemistry.

Introduction

In 1983, Nurrenbern and Pickering published a paper in the Journal of Chemical Education entitled, "Concept Learning versus Problem Solving: Is There a Difference?"[1]. The hypothesis of this paper was that students who could solve numerical problems also understood molecular concepts. Students were asked to solve traditional, algorithmic gas law questions, as well as conceptual problems about the behavior of gases that had no mathematical content. Students were also asked to answer another pair of algorithmic and conceptual questions about limiting reagents. The results indicated that students had significantly greater success solving the traditional, algorithmic problems than they did in correctly answering the conceptual, non-mathematical questions for both the gas law and the stoichiometry problems. The conclusion to this research was that "teaching students to solve problems about chemistry is not equivalent to teaching them about the nature of matter."

This research study raised far more questions than it answered. Why could students calculate the volume of a gas under changing conditions of temperature and pressure, but not identify a drawing of how the molecules of gas would be distributed in a steel cylinder of gas? How could students calculate the mass of product in a reaction as well as mass of unreacted reagents, yet not recognize a drawing that represented the number molecules of product and leftover reactants? What were chemistry faculty teaching students in the name of problem solving? How were students able to successfully answer the questions asked of them, yet struggle to answer related questions about the problems before them?

Nurrenbern and Pickering's study was not designed to examine these "whys" and "hows." Rather, the quantitative design of their research study provided convincing statistical evidence that a problem existed in the first place, i.e., that a difference between algorithmic and conceptual problem solving existed.

Exploring the origins of this difference necessitates a shift in experimental design. Illuminating how students think about gases and stoichiometry and discovering what mechanisms they employ to solve the problems requires collecting different kinds of data. Exploring these different kinds of questions requires different methodology and an altogether different research design. Answering these "why" and "how" questions requires an expertise in qualitative research methodologies.

Choosing Methodology: Qualitative vs. Quantitative

Which methodologies are more powerful? Should I develop a quantitative research design or a qualitative design? These are questions that typify an age-old debate in education research. Both quantitative and qualitative research designs have strengths. Both have limitations. Consider these representative questions that could serve as the thesis of a research study or a research proposal:

- Does *cooperative learning* improve *retention of students*?
- Do more students *enroll in organic chemistry* after taking general chemistry taught with *peer-led team learning*?
- Does a *guided inquiry* approach reduce student misconceptions about *thermochemistry*?

Each of these questions explores the relationship between two variables (in italics) involved in teaching and learning chemistry. The research model often employed to answer such questions strongly resembles the structure inherent in the scientific method: make hypotheses based on theory, design an experiment to test these hypotheses, gather data by manipulating the variable(s) of interest while controlling the others, analyze the results, revise the theory accordingly, make new hypotheses, and the cycle continues. How does this research design affect how chemists frame research problems in chemistry education? Consider this comment regarding an analogous situation in physics education research:

> "The basic strategy used by science educators to investigate these variables is to identify a single student characteristic (e.g., Piaget level) and demonstrate that the characteristic is correlated with success in physics; then, typically instruction is modified to take into account student inadequacies with respect to this characteristic, and studies are conducted to demonstrate that student achievement improves." (2, p. 61)

Certainly this research design is familiar to chemists and powerful for answering questions in the laboratory. However, this does not necessarily mean that this one research design is suitable for all research questions in chemistry education. Consider some of the questions identified in Chapter 2 by Zare (3) as important for chemistry education researchers to answer for chemistry departments:

- What academic structures cause undergraduates to major in chemistry?
- What is the importance of lecture demonstrations?
- How important are group learning activities?
- How should we handle beginning chemistry students with widely different backgrounds?

Clearly, these questions do not suggest an experimental design where one variable is manipulated while all others are held constant in order to confirm a hypothesis which defines the relationship between two such variables.

Chemistry education embraces a wide range of research questions; some are well suited to quantitative methodologies, while others are better answered by qualitative methods. Debating which is more powerful – quantitative or qualitative research methodologies – obscures a far more important point. One methodology is not inherently more powerful than the other. What is more important is to establish a good "fit" between the research questions and the research methodologies. The qualitative methodologies described in this chapter are well suited to exploring phenomenon that are not well understood or previously described, i.e., where formulating a working hypothesis would be difficult given the minimal knowledge in the literature regarding the important dimensions of the problem. This is not to say that a qualitative research study cannot explore the relationship between two specified variables; to the contrary, qualitative research often focus on examining relationships. What differs, though, from quantitative research designs is that a specific relationship is not defined *a priori* in the hypothesis.

In other settings, when the phenomenon of interest is well known, quantitative methodologies are much more powerful as researchers can now look for trends, examine correlations between variables, and consider under what conditions or for what groups of students these trends hold. These quantitative methodologies are the subject of Chapter 8 by Sanger (*4*). And at other times, an experimental design that utilizes both qualitative and quantitative methodologies is appropriate; in Chapter 9 (*5*), Towns describes the advantages of mixed-methods research designs that integrate both methodologies into one research project.

This chapter outlines important methodological considerations for research questions in chemistry education well suited to qualitative methodologies. Comparisons are drawn to key differences between qualitative and quantitative methodologies where possible.

Data Collection: Generating a Thick Description

Sampling

One familiar sampling strategy for chemists is that of drawing a random sample. The underlying premise of random samples is to avoid researcher bias, i.e., to collect data from a representative sample of the population of interest; sampling protocols other than random sampling are susceptible to suspicion that the researcher might skew the findings by selecting (consciously or subconsciously) subjects who validate the hypothesis.

Given that qualitative research designs rarely formulate a researchable hypothesis, this concern is irrelevant. Furthermore, qualitative research designs reject the postulate that there exists a distance of objectivity between the researcher and the research subjects. The case of Heisenberg's Uncertainty Principle is instructive for asserting the viability of this stance. In 1927, Heisenberg shook the scientific community by arguing that it was impossible to simultaneously determine both the momentum and the position of an electron. Heisenberg reported that in attempting to determine the position of the particle, the photon used to do so was itself responsible for altering that very position. As Heisenberg wrote (*6*), "what we observe is not nature itself, but nature exposed to our method of questioning."

In keeping with Heisenberg's findings, qualitative research designs operate from the premise that it is impossible to separate the inquirer from the subject of the inquiry because the two "interact to influence one another; knower and known are inseparable" (*7*). This interaction is, in fact, *valued* in qualitative research due to the multiple realities and meanings likely to be encountered by the human, who is uniquely capable of functioning as a research instrument in interviews and observations.

Consequently, in order to maximize this interaction between the researcher and the research subjects, most qualitative research designs utilize a **purposeful sampling** strategy (*8*) with information-rich cases to maximize the opportunity to produce a "**thick description**." (*9*) For example, suppose faculty wanted to design a research study to understand what factors caused students to drop organic chemistry. Rather than collecting data from a random sample of students enrolled in organic chemistry, a purposeful sample would be to collect data from those students who would provide the most information-rich data, i.e., students at greatest risk of dropping. Faculty should select a criterion to identify these students, e.g., performance in general chemistry.

Questions best answered through qualitative research designs typically place particular value upon the knowledge constructed (*10*) by individuals through their experiences and the meanings subscribed thereto. Therefore, qualitative methodologies must facilitate both the sharing of these multiple realities and the emergence of community constructions of meaning amongst all the research subjects. Qualitative methods are well suited for gaining entrée to the *emic (inside)* **perspective** of the research subjects, i.e., methods that allow the meanings of the participants to surface. By contrast, methods that would impose, *a priori*, theoretically or experimentally derived categories of meanings or experiences upon the research subjects risk stifling the emergence of the emic perspective and place higher priority upon validating already existing theoretical concepts, i.e., the *etic* (outside) **perspective**.

The most common methods of data collection in qualitative research designs are interviews, observations, and document analysis.

Human Subjects Research

Any research study involving human subjects must file an application with the local **Institutional Review Board (IRB)** for **Human Subjects Research**. The primary purpose of the IRB is to ensure that the rights of all human subjects are protected in accordance with federal regulations. Therefore, the first step in developing the research design for a project is to identify any and all dimensions of the project that will come under scrutiny by the IRB. Before collecting any data or recruiting any research subjects, the research design must be approved by the IRB.

The central feature of securing IRB approval is to ensure all human subjects provide **informed consent** for their participation. Typically, informed consent issues with students and teachers include ensuring the confidentiality of student responses, no coercion to participate in the study, secure storage of records, etc. Minors cannot give informed consent, so any study involving students under the age of 18 (which includes many college freshmen) must secure both parental consent and student assent. Research studies in K-12 settings require this plus approval of the local school board and/or district office.

Interviews

Gaining entreé to student thinking can be more fruitfully pursued through interviews than by student responses to a survey questionnaire with examples or categories pre-selected by the researcher. A key premise underlying qualitative research is to avoid placing constraints on the inquiry from the perspective of the researcher and to allow the categories to emerge from the data. Defining categories by which to sort participants before beginning the inquiry would undermine this premise. Rather, as Marshall and Rossman caution, "it is essential in the study of people to know just how those people define the situation in which they find themselves" (*11*).

Interviews typically employ a **semi-structured interview guide** that ensures all research participants are asked the same set of questions. The guide is semi-structured in that it contains prompts for the interviewer to request elaboration of additional details and examples in response to the personal views and ideas offered by the interviewee.

Interviews can be conducted one individual at a time, or with small groups of people, typically 4-6 people, in what are known as focus groups. Conducting an interview requires active listening skills; questions on an interview guide should not be asked in a mechanical manner with the goal of merely asking one after another. Skilled interviewers know how to encourage research subjects to think aloud (*12*), elaborate with examples, ask for contrasting or contradicting ideas, and weave together answers to separate questions to develop synthesis and analysis of emerging ideas in real time during the interview. Gubrium's *Handbook* offers a classic primer for conducting interviews (*13*). Interviews and focus groups are typically tape recorded and/or video recorded to be transcribed into a written record of the data.

Interviews played an essential role in advancing the research described at the start of this chapter. Using semi-structured interviews, Nakhleh (*14*) was able to go beyond showing that a difference existed between students who could calculate answers to questions but did not understand the properties of matter. Nakhleh's data characterized the thinking that students employ in solving these problems; she conducted think aloud interviews with students as they solved the conceptual and algorithmic problems in order to reveal how they made sense of the information in the problems.

Observations

Unlike quantitative designs that often try to remove the confounding effect of context from the variables of interest, qualitative research not only embraces such context, but actually considers the inclusion of context as essential to conducting the research. Because students in the same context, e.g., an instrumental analysis course, will interpret that experience differently, interviews may not be sufficient to understand the students' experiences. Observations provide a method for collecting data about the study's context in order to interpret the data gathered through interviews. Observations are crucial to understanding and documenting the significant influence of the setting in which students interpret their thoughts, feelings, and actions to construct meaning (*15*).

As with interviews, researchers in the field who collect observation data are typically highly skilled. Still, such researchers face choices in how to structure their research design. Should the researcher participate in the setting? For example, should a research study interested in understanding student learning in the lab utilize an observer who actually works at the bench and carries out the experiment along side the students, i.e., the research subjects? While doing so gives the researcher first-hand data as to the students' experience, the necessity of carrying out lab procedures necessarily limits the opportunity of the researcher to make notes and gather data about student actions and remarks during the lab experiment. Alternatively, the researcher may choose to be a non-participant, in order to enable a more complete collection of data for the study. Denzin and Lincoln offer important guidelines to consider when designing and conducting observations (*16*).

In addition to the choice of being a participant/non-participant, researchers also face the option of covert observation vs. overt observation. That is, does the researcher "go under cover" and not let students know of the research study and data collection efforts? In the lab study described above, a researcher could pretend to be a college student and work alongside "peers" or could secretly videotape students in the lab. The ethical dimensions of these problems require careful review by an Institutional Review Board for Human Subjects Research. Conversely, a research project can disclose the purpose of the research and fully inform students as to what data will be collected. The informants in Sheila Tobias' *They're Not Dumb, They're Different* (*17*) were overt participants in introductory college courses in chemistry and physics, keeping data logs regarding their experiences as they discussed what made learning difficult in these courses and the role of assessment in their learning.

These two dimensions of observation methodology are summarized in Table I.

Table I. Dimensions of Observation Methodologies.

Participation

	Overt Participant	Overt Non-Participant
Disclosure	Covert Participant	Covert Non-Participant

Document Analysis

The third most common method for collecting data in qualitative research designs is known as **document analysis**. While the thoughts, words, and actions of research subjects are essential, generating thick description requires a complete description of the context in which these thoughts and words are formed and actions take place. Course syllabi, copies of lab experiments, copies of student evaluation forms, and copies of research proposals are typical examples of documents gathered to analyze in a qualitative research design.

The methodology in Pienta's research (*18*) on productivity in chemistry education research used document analysis of records drawn from issues of the *ACS Directory of Graduate Research* (*19, 20*).

Triangulation

While any of these three research methods could be used singly in a research design, the **triangulation** (*8*) of these methods creates a more powerful design and provides analysis opportunities not available through the use of a single method. This marrying of methods is well established in chemistry as well. An organic chemist could use a GC-MS to separate and characterize the products of a synthesis, or may run both a ^{13}C NMR and obtain a crystal structure in order to gather information from multiple perspectives about the compound of interest. So it is with triangulation and qualitative research methods.

For example, a researcher may observe the lectures in a general chemistry course and repeatedly hear the instructor characterize the laboratory experiments as inquiry-oriented. Yet, a document analysis of the actual experimental procedures using a rubric for laboratory inquiry reveals that most of the experiments require the students to do little more than follow directions and perform calculations as directed. Or, during interviews students may tell a researcher that they consider themselves to be hands-on learners and that visual information is very important for their learning. Yet, during observations of homework problem-solving sessions throughout the semester, the researcher observes that these same students do not once pick up a model kit and build models of compounds to help them answer the homework questions even when such tools are readily available.

In these cases, triangulation can reveal inconsistencies or contradictions in the data. Triangulation can also function to provide complementary data as well. Researchers may find themselves frustrated by students who are not particularly articulate during an interview about their problem-solving strategies. However, by including multiple data collection methods in the research design, this same researcher can glean important data during observations of these same students (e.g., as they solve problems in class) that was unavailable during the interviews.

Fieldnotes

Just as chemists make regular entries in their lab notebooks about procedures, data, and emerging hypotheses, so too do qualitative researchers. Qualitative research designs rely heavily on accurate and methodical record-keeping. **Fieldnotes** are created for each data collection opportunity and typically include the following (*21*):

Log

The log section of fieldnotes includes the factual information regarding the date and site of data collection, the type of data collection (covert participant observation, focus group interview, document analysis of student evaluations, etc.), the names of both the research subjects and the researchers collecting the data, and a statement of the purpose of this data collection activity.

Data

This section of the fieldnotes log includes the data themselves such as a transcript of the interview or notes from the observation. (*N.B.* Transcription is a time-consuming and expensive process, often requiring 3-4 hours of transcription for every 1 hour of interview.)

Analytical Comments

After conducting the first individual interview or observing only one class, qualitative researchers should begin immediately to identify possible themes, ideas, or working hypotheses that emerge from the data collection. Certainly, these themes will be speculative in the initial data collection activities. As data collection continues, however, deeper insights and thoughtful reflections will develop that either support speculative hypotheses or refute emerging themes from earlier data collection activities.

Methodological Comments

Fieldnotes should include not only analytical comments about emerging findings from the data collection, but should also include reflections from the researcher concerning his/her impressions about the quality of the data collected. Typical comments in this section include changes to make in the semi-structured interview guide given any difficulties that arose, thoughts about additional information-rich cases to interview based on what was learned in this data collection, modified strategies for future observations, etc. This ability to make modifications and improvements in the methodologies in real-time as the study progresses is a hallmark of qualitative research known as **emergent design**.

Selecting a Qualitative Research Tradition

Qualitative research is not unique to chemistry education, nor does its origins lie within chemistry. Rather, qualitative research traditions have developed among multiple social science disciplines such as anthropology, psychology, and sociology. Chemistry education adopts these research traditions and adapts them to understanding the questions of teaching and learning of chemistry. A comprehensive discussion of all qualitative research traditions, or even the primary ones, is beyond the scope of this chapter; readers are referred to Erickson (*22*), Roth (*23*), Creswell (*24*), Merriam (*25*), and Guba and Lincoln

(*26*) for detailed descriptions of each tradition. A brief introduction to commonly used traditions in chemistry education is provided below, including representative research questions well suited to each tradition.

Case Study

As the name suggests, the **case study** involves a detailed examination of an individual case, which in chemistry education research is most likely to be an individual student or an individual instructor. In order to gather sufficiently detailed data about both the individual and the context(s) in which the boundaries of the case exist, the researcher employs multiple sources of data and multiple methods to triangulate the findings.

Lyons, Freitag, and Hewson (*27*) used a case study methodology to illuminate the alignment (or inconsistencies) between espoused theories of learning and practices in the classroom for an experienced high school chemistry teacher, Mr. Corrigan. Through the use of interviews and classroom observations, the authors wrote a case study for Mr. Corrigan which contrasted his priority on students learning to think and explore in the laboratory with his classroom management practices which stifled such exploratory thinking.

Sweeney, Bula, and Cornett's study (*28*) of a beginning high school chemistry teacher used a case study methodology to understand the role of reflection upon the teacher's development and classroom practice. An additional example of case study methodology in chemistry education research can be found in Harrison and Treagust's investigation (*29*) of high school students' use of models in understanding chemistry concepts.

Phenomenology

Phenomenology is a tradition which focuses on understanding the direct experiences of participants in a given context and the value those experiences have for the participants. The structure of the phenomenon and its identifying characteristics are also essential to discern in a phenomenological inquiry.

For example, Koballa and colleagues investigated pre-service high school chemistry teachers' conceptions of teaching and learning at a German University (*30*). Their two research questions were

- How do pre-service chemistry teachers conceptualize chemistry learning and teaching during their undergraduate experience?
- What relationships exist between pre-service teachers' conceptions of learning chemistry and teaching chemistry?

The findings of this research showed that pre-service teachers viewed learning chemistry as the ability to reproduce knowledge given by the teacher, while they viewed teaching chemistry as the ability to facilitate that reproduction of knowledge.

Koballa and colleagues were particularly interested in identifying the origins of these conceptions by the pre-service teachers as they related to the experiences of these pre-service teachers as undergraduates in chemistry classrooms. Selecting a phenomenographical methodology to frame the study enabled the researchers to examine multiple dimensions of the phenomenon of being an undergraduate chemistry student. In particular, Koball and colleagues examined how these experiences in undergraduate chemistry extended into the future, i.e., how undergraduate pre-service teachers' own struggles and successes with learning chemistry now shaped their own ideas about teaching chemistry and what to expect from their students as learners of chemistry.

Other representative phenomenological studies in chemistry education include Ebenezer's studies of solubility (*31*) and solution formation (*32*), Orgill and Bodner's research regarding the use of analogies in teaching biochemistry (*33*), and Liu and Lesniak's investigation of how children's understandings about matter change from elementary school to high school (*34*).

Ethnography

Ethnography of cultures, social beliefs, values, and traditions is the methodology that describes much research in anthropology. As applied to chemistry education research, ethnographic studies are frequently interested in understanding the culture of a particular classroom, e.g., an AP chemistry classroom, or a prep chem class for underprepared students at the university. Ethnography focuses on identifying the social norms that regulate behavior and community knowledge production. In a chemistry research laboratory, social norms might encourage students to share data regarding related projects. Conversely, however, for students in a general chemistry laboratory course, the culture of competition or the professor's practice of grading on the curve may discourage collaboration.

Rop's research study (*35*) sought to answer these two questions:

- What does it mean to be successful in introductory chemistry?
- What does it mean to understand chemistry?

He conducted this research as a participant observer for one year in two chemistry classes at a high school in the midwestern United States. The findings of this research not only include answers to the above questions, but also identify the "spheres of influence" within the classroom that shape the cultural norm of success in a high school chemistry classroom.

Other ethnographic chemistry/science education research studies include Driver's classic studies on student misconceptions (*36*), Barton's studies on science education in urban settings with homeless children (*37*), and Hammond's research on bilingual science education (*38*).

Grounded Theory

The tradition of **grounded theory** (*39*) has its roots in sociology. Grounded theory is a theoretical framework that emerges from the iterative analysis of data based on coding and category building. The ultimate purpose of building this theory is to test its validity against other cases.

Taber's paper (*40*) provides an excellent, detailed description of grounded theory for chemists as well as reports on his selection of this qualitative tradition to frame his inquiry into students' developmental understanding of the concept of the chemical bond. In contrast to other qualitative research traditions, grounded theory studies typically do not frame a research question to be answered, but rather begin by identifying a concern situated in a real context that warrants closer examination. In Taber's case, he had noticed that students could articulate satisfactory explanations of chemical bonding at the secondary level, but that these same students struggled to develop more sophisticated models of bonding at the university. Taber's data consisted of student explanations regarding models of atoms, molecules, and ions, as well as their understanding of concepts such as chemical stability and ionization energy. Merely coding the students' descriptions of these individual concepts would have provided little insight into student thinking. The processes of coding and seeking relationships between the codes in grounded theory brought the "bigger picture" to the forefront and offered a wholistic, integrated view of student thinking about chemical bonding.

Data Analysis

"What we call our data are really our own constructions of other people's constructions of what they and their compatriots are up to ... analysis, then, is sorting out the structures of signification." (*9*)

This quote from Geertz's *The Interpretation of Culture* points out that data collected using qualitative methodologies, e.g., the content of the interviews and the transcripts that visualize this data, are constructions offered by the participants. That is, this data is a representation of how the research subjects make sense of their experiences; during an interview they are sharing information about their own meaning making processes. Data analysis then, in the context of these qualitative methodologies, becomes the process of the researcher formulating yet another set of constructions, i.e., *re-constructions*, (*7*) of research subjects' own constructions.

Theoretical Framework

Qualitative research designs rarely stipulate a researchable hypothesis because of the value placed upon understanding the experiences of the research subjects and the meanings they attributed to their experiences. Qualitative research has a reputation as "soft-science" or that it lacks grounding in a theoretical framework. This common misconception leads to the perception that qualitative research designs are "theory-free."

To the contrary, well-designed qualitative research studies always stipulate the theoretical underpinnings that shape the inquiry. Chapter 5 by Abraham (*41*) articulates the important issues in selecting a theoretical framework. In chemistry education research, these theoretical frameworks typically articulate the principles of how learning takes place, what barriers prevent learning from occurring, and how these principles can be put into practice. In 2001, the *Journal of Chemical Education* published a symposium (*42-49*) highlighting several dominant theories of learning that inform the teaching and learning of chemistry and research about these processes.

Coding and Categories

Although the specific methods of analysis vary somewhat amongst the different qualitative traditions, most qualitative research designs include the concepts of **coding** data and inductive data analysis to develop **categories**. For example, after interviews are transcribed verbatim, text is segmented into phrases or sentences and given a code that labels the main idea of this "chunk" of text.

This process of "chunking data" and developing codes is an iterative process as each new data collection opportunity informs past codes as both similarities and refined differences in meanings emerge from the data. Related codes are then grouped into categories, which is itself another iterative process, until a self-consistent category system emerges from the data. All iterations of codes and categories are carefully noted in the analytical comments of fieldnotes to facilitate an audit trail (see below). Several commercial software packages are available to assist with coding and categorization, such as NVivo (*50*), Ethnograph (*51*), Atlas/ti (*52*), and HyperRESEARCH (*53*).

Data Quality

All research claims are ultimately subject to questions of **reliability** and **validity**. How do we know that the findings are "real" and did not occur by chance? How do we know that these findings might apply to other students in other classes? How do we know that the findings are free from researcher bias? Such are the questions that shape the critically important process of peer review in judging the quality of data collected in a research study.

With regard to researcher bias, we have already discussed the rejection of maintaining an objective distance between the researcher and the researched – qualitative research views the human being as uniquely capable of crossing this gap and developing understandings of fellow human experiences and the meanings they attach to such experiences. Furthermore, most qualitative research designs do not involve "one-shot" data collections, but rather involve **prolonged engagement** of the researcher in the setting of interest, thereby reducing any temporary effects of introducing an outsider into the research setting.

Not surprisingly, the concepts of reliability and validity do not seamlessly juxtapose with most qualitative research methods. In *Naturalistic Inquiry*, Lincoln and Guba (*7*) articulated trustworthiness criteria to defend the knowledge claims made in qualitative inquiry. These criteria focus on the **credibility, transferability, dependability,** and **confirmability** of the emic perspective as discerned through qualitative research methods.

Transferability and Thick Description

The extent to which qualitative research findings are transferable from one chemistry classroom to another depends on the similarities between the context of the research and the other context under consideration. A qualitative researcher cannot anticipate all such contexts to which transferability might be sought. Such knowledge is held only by the reader. Accordingly, the responsibilities of the researcher in ensuring transferability are fulfilled by providing the "**thick description**" of the context of the inquiry in order to best facilitate comparisons made by readers to their own contexts of interest. By describing in detail the constructions and experiences of the participants and the circumstances of the context in which such meanings were constructed, the readers of a qualitative research study will be able to recognize both similarities to and important differences from their own contexts of interest.

The issue of **transferability** highlights an importance difference between chemistry research at the bench and chemistry research conducted about student learning as described in a report by the ACS Division of Chemical Education's Task Force on Chemistry Education Research:

> *"The subject matter of chemistry tends to hold still, making mathematical description somewhat easier than in the case of chemistry education. Chemists operate under the assumption that a collection of hydrogen molecules today is indistinguishable from one assembled 50 years ago. However, student bodies change from semester to semester, and students are exposed to countless influences that are difficult to describe as mathematical variables."* (54, p. 851)

Therefore, qualitative researchers in chemistry education are cautioned to report their findings with detailed contextual information, keeping in mind that such a system constantly morphs while you are trying to measure it. This will place boundary conditions upon the transferability of the research to other chemistry classrooms.

Credibility and Member Checks

Issues of credibility, i.e., concern that the researcher's re-constructions and knowledge claims are "credible to the constructors of the original multiple realities," (7, p. 296) can be addressed through a procedure known as a **member check**. Member checks regularly take place during all interviews by the researcher taking care to summarize and reflect back to the interviewee(s) the researcher's understanding of their remarks. Member checks can also be used during the analysis to identify working hypotheses or emerging assertions. For example, in addition to transcribing interviews, a concept map (55-61) can be prepared by the researcher to summarize the key ideas and examples offered by the interviewee. This concept map can both summarize the interview's content as well as represent the researcher's interpretations. After briefly explaining the structures inherent in a concept map (i.e., concepts and propositions), the concept map can be shared with the interviewee for discussion and feedback. The interviewee should review the concept map for accuracy and completeness. Specific probes the researcher can ask during this process include "what concepts or ideas did we talk about that are missing from this concept map?" or "what concepts are not linked, but should be?," or "what linking words would you change to better represent your ideas?"

Dependability, Confirmability, and Audit Trails

Dependability concerns the quality of the inquiry *process* while **confirmability** addresses the quality of the inquiry *product*, i.e., the data, interpretations and recommendations. A common mechanism to address both of these trustworthiness criteria is to employ a qualitative data audit in the research design. An **audit trail** can simultaneously address both the dependability and the confirmability of a research study. A dependability audit reviews the methodological decisions made in the inquiry, while the confirmability audit examines the findings of the inquiry to confirm that the researcher's interpretations are grounded in both the theoretical framework and in the data collected.

Audit trails consist of all possible forms of documentation about the process and products of the research. For example, an audit trail would typically include a copy of the original research proposal, interview guides, transcripts, concept maps, fieldnotes (including research memos articulating methodological decisions and analytical developments), category systems, and a draft of the study's findings. These documents are then given to colleagues who are independent of the project, but skilled in qualitative research. These auditors (typically two) review the project both in chronological order and in a holistic manner, testing the documents to see if they validate the assertions/working hypotheses which emerged from the research; auditors are also charged to conduct a negative case analysis, i.e., to see if they can find data which negates or refutes the emerging hypotheses. Lincoln and Guba (7) and Halpern (62) offered detailed guidelines and questions that auditors might ask of the materials in the audit trail.

Qualitative Research in a Chemistry Department: Practical Considerations

Qualitative research is not the predominant research design in chemistry departments, nor even in chemistry education research. Consider this statistic: in the last decade, the *Journal of Chemical Education* has published over 310 research studies, only 8 of which utilized a qualitative research design. Because chemistry education research utilizing a qualitative research design is unfamiliar to many chemists, researchers need to be aware that colleagues may bring unrealistic expectations regarding outcomes of the research to peer review processes such as manuscript reviews or evaluation of promotion dossiers. Tobias offers this caution in *Revitalizing Undergraduate Science*:

"Trained in problem definition and problem solving, scientists inevitably bring the habits of doing science to the problem of [science

education] reform ... First, they believe there is one best curriculum or pedagogy waiting to be discovered, like the laws of nature, like quarks. If it hasn't been discovered so far, it's because researchers haven't worked hard enough. This idealized curriculum or pedagogy is not only 'right,' it is universal, and will work best irrespective of teacher, content, and place. Second, by pursuing abstract studies of the nature of knowledge and cognition, researchers can find this curriculum or pedagogy and experimentally prove it is the best. And third, such experimental evidence will persuade instructors everywhere to adopt the program." (*63*, p. 16)

Chemists who conduct qualitative research should recognize that a large segment of their target audience may expect the results of such research to prove that a particular pedagogy or curriculum is "best" for students. Failure of the research to make such a claim or to collect data in support of making such a claim may be perceived as the result of a flawed research design. Chemistry education researchers need to be proactive in communicating the underlying philosophy and the value of the outcomes of their research.

As chemistry education research studies are designed, and the question arises as to what methodologies will frame the inquiry, Dennis Jacobs, Carnegie Scholar and Professor of Chemistry at University of Notre Dame (*64*) advises researchers to remember that the research question must be credible:

- Who is the audience for your findings? What will your audience perceive as the most credible evidence?
- What kinds of evidence will you gather? Will you gather data about the *outcome of a course* or data about *what's happening in the course*?

Summary

In qualitative research, it is incumbent upon the researcher to elicit not only the meanings that individuals hold within a given context, but also the experiences and feelings they ascribe to such constructions of meaning. Lincoln and Guba (*7*) offer fourteen guidelines, which they refer to as the "characteristics of operational naturalistic inquiry": natural setting, human instrument, utilization of tacit knowledge, qualitative methods, purposeful sampling, inductive data analysis, grounded theory, emergent design, negotiated outcomes, case study reporting mode, idiographic interpretation, tentative application, focus-determined boundaries, and special criteria for trustworthiness. Each of these salient features must be carefully considered and articulated to fit with the research question of interest in a qualitative research design.

Recommended Readings

- *Qualitative Methodologies in Chemical Education Research* by Amy Phelps, *Journal of Chemical Education*, **1994**, *71(3)*, 191.
- *The Use of Triangulation Methods in Qualitative Educational Research* by Maria Oliver-Hoyo and DeeDee Allen, *Journal of College Science Teaching*, **2006**, *35(4)*, 42.

References

1. Nurrenbern, S.C.; Pickering, M. *J. Chem. Educ.*, **1983**, *64*, 508.
2. Champagne, A. B.; Gunstone, R. R.; Klopfer, L. E. In *Cognitive Structure and Conceptual Change*; L. H. T. West and A. L. Pines (Eds.); Orlando, FL: Academic Press, 1985.
3. Zare, R.N. In *Nuts and Bolts of Chemical Education Research*; D. Bunce and R. Cole (Eds.); American Chemical Society Symposium Series, Washington, D.C.: 2007.
4. Sanger, M.J. In *Nuts and Bolts of Chemical Education Research*; D. Bunce and R. Cole (Eds.); American Chemical Society Symposium Series, Washington, D.C.: 2007.
5. Towns, M.H. In *Nuts and Bolts of Chemical Education Research*; D. Bunce and R. Cole (Eds.); American Chemical Society Symposium Series, Washington, D.C.: 2007.
6. Heisenberg, W. *Physics and Philosophy*. New York: Harper and Row. 1958.
7. Lincoln, Y.S.; Guba, E.G. *Naturalistic Inquiry*. Beverly Hills, CA: Sage. 1985.
8. Patton, M. Q. *Qualitative Evaluation and Research Methods*. Newbury Park, CA: Sage. 2002.
9. Geertz, C. *The Interpretation of Cultures: Selected Essays*. New York: Basic Books. 1973.
10. Bodner, G.M. *J. Chem. Educ.*, **1986**, *63*, 873.
11. Marshall, C.; Rossman, G. B. *Designing Qualitative Research*. Newbury Park, CA: Sage. 1989.
12. Bowen, C.W. *J. Chem. Educ.*, **1994**, *71*, 184.
13. Gubrium, J. *Handbook of Interview Research*. Newbury Park, CA: Sage. 2001.
14. Nakhleh, M.B. *J. Chem. Educ.*, **1993**, *70*, 52.
15. Wilson, S. *Rev. Educ. Res.*, **1977**, *47*, 245.
16. Denzin, N.K. and Lincoln, Y.S. *The Sage Handbook of Qualitative Research*. Newbury Park: Sage. 2005.

17. Tobias, S. *They're Not Dumb, They're Different: Stalking the Second Tier.* Tucson, AZ: Research Corporation. 1990.
18. Pienta, N. *J. Chem. Educ.*, **2004**, *81(4)*, 579
19. *ACS Directory of Graduate Research.* American Chemical Society: Washington, DC. 1997.
20. *ACS Directory of Graduate Research.* American Chemical Society: Washington, DC. 1999.
21. Greene, J. **1993**. Personal Communication.
22. Erickson, F. in *International Handbook of Science Education, Vol. II.* Fraser, B.J.; Tobin, K.G. (Eds.); Kluwer Academic Publishers: London. 1998.
23. Roth, W.-M. *Doing Qualitative Research: Praxis of Method.* Sense Publishers: Rotterdam, 2005.
24. Creswell, J.W. *Qualitative Inquiry and Research Design: Choosing from among Five Traditions.* Sage Publications: Thousand Oaks, 1998.
25. Merriam, S.B. *Qualitative Research and Case Study Applications in Education.* Jossey-Bass: San Francisco, 1998.
26. Guba, E. G.; Lincoln, Y. S. *Fourth Generation Evaluation.* Newbury Park, CA: Sage, 1989.
27. Lyons, L.L.; Freitag, P.K.; Hewson, P.W. *J. Res. Sci. Tchg.* **1997**, *34(3)*, 239.
28. Sweeney, A.E.; Bula, O.A.; Cornett, J.W.; *J. Res. Sci. Tchg.*, **2001**, *38(4)*, 408.
29. Harrison, A.G.; Treagust, D.F. *Sci. Educ.*, **2000**, *84 (3)*, 352.
30. Koballa, *Intl. J. Sci. Educ.* **2000**, *22(5)* 469.
31. Ebenezer, J. *Sci. Educ.*, **1996**, *80(2)*, 181.
32. Ebenzer J.; Gaskell, J. *Sci. Educ.*, **1996**, *79(1)*, 1.
33. Orgill, M, Bodner, G. *Chem. Educ. Res. Practice*, **2004**, *5*, 15.
34. Liu, X; Lesniak, K. *J. Res. Sci. Tchg.*, **2006**, *43(3)*, 320.
35. Rop, C.J. *J. Res. Sci. Tchg.*, **1999**, *36(2)*, 221.
36. Driver, R. *Intl. J. Sci. Educ.*, **1989**, *11*, 481.
37. Barton, A.C. *J. Res. Sci. Tchg.*, **2001**, *38(8)*, 899.
38. Hammond, L. *J. Res. Sci. Tchg.*, **2001**, *38(9)*, 983.
39. Strauss, A.; Corbin, J. *Basics of Qualitative Research: Techniques and Procedures for Developing Grounded Theory.* Sage Publications: Thousand Oaks, 1998.
40. Taber *Intl. J. Sci. Educ.*, **2000**, *22(5)*, 469.
41. Abraham, M.R. In *Nuts and Bolts of Chemical Education Research*; D. Bunce and R. Cole (Eds.); American Chemical Society Symposium Series, Washington, D.C.: 2007.
42. Nurrenbern, S.C. *J. Chem. Educ.*, **2001**, *78(8)*, 1107.
43. Bunce, D.M. *J. Chem. Educ.*, **2001**, *78(8)*, 1107.
44. Samarapungavan, A.; Robinson, W.R. *J. Chem. Educ.*, **2001**, *78(8)*, 1107.

45. Bodner, G.M.; Klobuchar, M.; Geelan, D. *J. Chem. Educ.*, **2001**, *78(8)*, 1107.
46. Nakhleh, M.B. *J. Chem. Educ.*, **2001**, *78(8)*, 1107.
47. Bretz, S.L. *J. Chem. Educ.*, **2001**, *78(8)*, 1107.
48. Towns, M.H. *J. Chem. Educ.*, **2001**, *78(8)*, 1107.
49. Wink, D.J. *J. Chem. Educ.*, **2001**, *78(8)*, 1107.
50. QSR International, URL http://www.qsrinternational.com/. (accessed October 2006)
51. Qualis Research Associates, The Ethnograph, URL http://www.qualisresearch.com/. (accessed October 2006)
52. ATLAS.ti, URL http://www.atlasti.com/. (accessed October 2006)
53. ResearchWare, URL http://www.researchware.com. (accessed October 2006)
54. Bunce, D.; Gabel, D; , Heron, D; Jones, L. *J. Chem. Educ.*, **1994**, *71(10)*, 850.
55. Cardellini, L. *J. Chem. Educ.*, **2004**, *81*, 1303.
56. Francisco, J.S.; Nakhleh, M.B.; Nurrenbern, S.C.; Miller, M.L. *J. Chem. Educ.*, **2002**, *79*, 248.
57. Nicoll, G; Francisco, J.S.; Nakhleh, M.B. *J. Chem. Educ.*, **2001**, *78(8)*, 1111.
58. Robinson, W.R. *J. Chem. Educ.*, **1999**, *76*, 1179.
59. Regis, Al; Albertazzi, P.G.; Roletto, E. *J. Chem. Educ.*, **1996**, *73*, 1084.
60. Pendley, B.D.; Bretz, R.L.; Novak, J.D. *J. Chem. Educ.*, **1994**, *71*, 9.
61. Stensvold, M.; Wilson, J.T. *J. Chem. Educ.*, **1992**, *69*, 230.
62. Halpern, E. S. *Auditing Naturalistic Inquiries: The Development and Application of a Model.* Unpublished doctoral dissertation, Indiana University, Bloomington, IN, 1983.
63. Tobias, S. *Revitalizing Undergraduate Science.* Research Corporation, Tucson, AZ, 1992.
64. Jacobs, D. Conference on Scholarship of Teaching and Learning, Youngstown State University, Youngstown, OH, 2003.

Chapter 8

Using Inferential Statistics to Answer Quantitative Chemical Education Research Questions

Michael J. Sanger

Department of Chemistry, Middle Tennessee State University, Murfreesboro, TN 37132

While many chemists value quantitative, statistical research more than other chemical education research involving less mathematical and more qualitative methodologies, few are comfortable performing or evaluating this kind of research. This chapter outlines eight steps for performing or evaluating chemical educational research involving inferential statistics. In each step, several common statistical terms are defined and described. This chapter also describes misconceptions demonstrated by novice chemical educational researchers, and explains the scientifically-accepted conceptions related to these misconceptions. In addition, this chapter uses examples from the chemical education literature on visualization techniques (static pictures, computer animations, etc.) to demonstrate how chemical education researchers decide which statistical tests (including t-tests, ANOVAs and ANCOVAs, statistical tests of proportions, and the non-parametric chi-square tests) to use in evaluating the research question posed in these studies.

© 2008 American Chemical Society

Introduction

As new subdisciplines of chemistry grow and develop, some are more easily accepted than others by mainstream chemists. Traditionally, chemical education has not been accepted as a true subdiscipline in chemistry. One reason may be that chemists don't view chemical education as a field where research leads to important new information (many don't know that chemical educators even do research). Another more subtle reason is that when chemical education researchers perform research, they are often using very different research methodologies that traditional chemists don't understand. In general, chemists tend to be more comfortable with the results of quantitative (statistical) research methodologies than the more qualitative methods described in Bretz's chapter of this book (*1*). Even though chemists may be more comfortable with quantitative methods, many do not understand the general concepts behind these statistical methods or their use in chemical education research. As an example, when making comparisons among students to answer a research question, many novice chemical education researchers do not report any statistical data regarding the students' performance or simply report descriptive statistics (means and standard deviations) and expect the readers to make conclusions based on that data. Without statistical comparisons, it is impossible to determine whether any reported values are different.

The purpose of this chapter is to provide guidance to those who are not familiar with quantitative chemical education research methodologies. This chapter outlines a series of steps that can be used to systematically answer quantitative chemical education research questions using inferential statistics. While the specific steps used in hypothesis testing vary from book to book, there is a good consensus regarding the steps required to perform statistical analyses of chemical education data (*2, 3*). Throughout this chapter, common misconceptions or mistakes regarding inferential statistics are described along with the more correct conceptions or statements that should be used. This chapter also discusses statistical studies performed by chemical education researchers on the topic of visualization techniques as a way to illustrate how chemical education researchers select appropriate test statistics to analyze their data and answer their research questions.

Step 1: State the Research Hypothesis

Bunce (*4*) describes the issues associated with writing a research question in greater detail. In this chapter, a distinction between research questions (which are often large and general) and research hypotheses (which are specific questions related to a statistical analysis) is made. Often, research questions are answered in parts using several specific research hypotheses (if the questions lend themselves to statistical analysis), qualitative research methods, or a combination of the two. Chemical education research studies involving qualitative and quantitative methods (mixed methods) are discussed in Town's chapter in this book (5). A common difficulty novice chemical education researchers face is turning their research questions into measurable research hypotheses. Research questions tend to use vague language (like 'improve' or 'understand') without specifically describing how these terms will be turned into measurable events or data to answer specific research hypotheses.

Chemical education researchers should note that how research questions are worded will affect how the studies are designed. For example, if the researchers were interested in determining whether a new instructional method would lead to student learning, responses to the same (or similar) questions given before and after instruction would be compared for the same set of students. If the researchers were interested to investigate whether a new instructional method would be an acceptable alternative to another instructional method, then one group should receive instruction using the existing method and the other should receive instruction using the new method, and the responses from these groups would be compared. If the researchers were interested in determining whether a new instructional method would be a useful supplement to an existing instructional method, then one group should receive instruction using the existing method and the other group should receive instruction using both methods, and the responses from the groups would be compared.

One way to illustrate the conversion of research questions into research hypotheses is to look at a specific example from the chemical education literature. Sanger, Phelps, and Fienhold (6) investigated whether a computer animation of a can crushing demonstration would improve students' conceptual understanding of the chemical processes occurring in the demonstration. Their research question was: "Does the computer animation improve students' conceptual understanding?" The phrase 'improve conceptual understanding' is vague. What does that mean? And more importantly, how is it measured?

In order to answer the research question, they needed to determine how they could convince themselves that students understood the demonstration concepts. This led to three specific research hypotheses: (1) "Did viewing the animation change the proportion of students correctly predicting that the can would collapse when cooled?" (2) "Did viewing the animation change the proportion of students blindly quoting the ideal gas laws (an indicator that they did not understand the concepts)?" (3) "Did viewing the animation change the proportion of students mentioning important concepts like the condensation of water or the decreased pressure inside the container?" Each of these ideas is now measurable and also more objective, since it is much easier for two individuals to agree on these specific hypotheses than the research question.

Chemical education research is most likely to provide valuable information to researchers when it is based on a specific theory. There are many reasons why literature reviews are particularly important in planning **theory-based research studies** (7, 8). Without a literature review, researchers have no idea what studies have already been performed, and what evidence already exists to support or contradict their research hypotheses. The literature review can also provide suggestions of possible research hypotheses that have not yet been answered. Although there are many relevant and useful research studies in the chemical education literature that are not theory-based, having a theory-base usually improves chemical education research studies. If the researchers are studying an area that is relatively new or unexplored, there may not be many relevant chemical education research studies. In this case, the researchers may have to expand their literature search to new areas (science education, psychology, or other areas) to find relevant studies.

Misconception—Directional research (and null) hypotheses should be used extensively: Most research hypotheses are interested in looking for differences between two or more groups. These differences can be among different groups of students experiencing different instructional strategies or from the same group of students before and after they have viewed an instructional lesson. While it is tempting for researchers to put directionalities in their research hypothesis (i.e., asking whether an instructional method will cause students will perform *better* rather than *differently* than students who did not receive it), from a statistical standpoint it is generally considered to be improper, even when based on educational theories or existing empirical research studies (3). While directional research hypotheses have their place, chemical education researchers should refrain from using them unless there is a specific reason to do so.

One major concern with directional hypotheses is that it eliminates some of the objectivity of a study. While researchers may have reason to believe that an instructional treatment will have a positive effect on student learning (especially if their research hypotheses are supported by existing chemical education research or theories), they should be open to the possibility that the instructional method could have *negative* effects on student learning. While (as instructors) researchers may believe that an instructional method should improve student learning, they cannot always predict whether the method will have the intended effect on students and it is always possible that the treatment might actually *distract* students from learning or may even *confuse* them. Another more serious concern regarding directional hypotheses is that by ignoring the possibility of the instructional effect being either positive or negative, the chemical education researcher can "overestimate" a particular effect (positive or negative). In other words, when posing directional research hypotheses, researchers are much more likely to conclude that there is an effect when, in fact, there is none.

Step 2: State the Null Hypothesis

Misconception—It is possible to prove that a research hypothesis is correct: This misconception usually appears when chemical education researchers say that they have proven their research hypotheses were correct. As chemists, we know that it is impossible to prove a theory correct, it can only be proven incorrect. If the data and theory do not agree, then we can be confident that the theory is wrong, assuming the data were properly collected. However, if the theory and data do agree, that does not necessarily mean that the theory is correct, or that it is the only theory that could be correct. As a result, our descriptions in this case are usually more tentative, and we say that the theory is consistent with or supported by the data and *could* be correct.

Inferential statistics are done much the same. Instead of trying to prove that their research hypotheses are correct (an impossible task), chemical education researchers approach the problem a different way—by trying to prove that the opposite hypotheses (also called **null hypotheses**) are incorrect. Since they are usually looking for the effect (whether positive or negative) of a treatment on student scores, the null hypotheses that are actually tested are that there are *no* effects due to the treatments. Once the test statistics (t, z, F, χ^2, etc.) are calculated, they are evaluated with respect to the *null* hypothesis. If the calculated statistics are shown to be very unlikely were the null hypotheses true, then the null hypotheses can be rejected and the research hypotheses are supported by the data. If the calculated statistics are not unlikely given the null hypotheses, then the researchers fail to reject the null hypotheses and the research hypotheses are not supported by the data. This does not necessarily mean the research hypotheses are wrong, it simply means that the data collected did not support the research hypotheses. An analogy from the American jurisprudence (legal) system can explain these differences. Failing to reject a null hypothesis would be analogous to a jury not finding a plaintiff guilty of a particular crime. However, not finding a plaintiff guilty is *not* the same as finding them innocent; it simply means that there was not enough evidence to convince the jury that the person was guilty beyond a reasonable doubt.

In the example described above (*6*), the null hypotheses would be: (1) Viewing the animation had no effect on the number of students correctly predicting that the can would collapse when cooled, (2) Viewing the animation had no effect on the number of students blindly quoting the ideal gas laws, and (3) Viewing the animation had no effect on the number of students mentioning important concepts like water vapor condensation or a decreased inner pressure.

Misconception—Hypothesis testing provides absolute proof that a research hypothesis is correct or incorrect: Many novice chemical education researchers don't recognize that results of hypothesis testing using inferential statistics are tentative and based on probabilities, not certainties. This stems from a confusion between descriptive and inferential statistics. **Descriptive statistics** are parameters used to *describe* a population of students, and include population means, medians, modes, standard deviations, and variances to name a few. Each statistical parameter represents a summary (simplification) of all scores within the population into a single value. Descriptive statistics are exact values, and have no error associated with them. The average scores for a group of students are what you calculate them to be, and chemical education researchers are infinitely confident that these values are correct for this population of students.

Inferential statistics, on the other hand, are parameters used to make *inferences* (generalizations) from a *sample* of the population to the population as a whole. Inferential statistics make use of descriptive statistics (means, standard deviations, etc.) of the sample to answer hypotheses about the entire population. Inferential statistical methods make the fundamental assumption that the sample from the population is representative of the whole population. Because this may or may not be true, inferential statistics are not exact values and there is always some probability of error associated with these values. While chemical education researchers collect data using students enrolled in a particular section of a course taught at their school in their region of the country, they would like to generalize the results of their research to a larger population, whether it is to all students enrolled in this particular course at their school or to all students at all levels in this country and abroad. Unfortunately, as the target population becomes bigger, it is harder to be confident that the sample of students used in this particular study is representative of the population as a whole and therefore the conclusions based on these data become more tentative.

Step 3: Define the Level of Significance

There are two types of errors that could occur during hypothesis testing using inferential statistics. The first is that a chemical education researcher rejects the null hypothesis when it is really true (called a **Type I error** or α); the second is that a researcher fails to reject a null hypothesis that is false (called a **Type II error** or β). When Type I errors are made, researchers conclude that there is an effect due to the instructional method when in reality there is not; when Type II errors are made, researchers fail to observe an instructional effect that really exists. While researchers would really like to avoid making Type II errors, making Type I errors are much more serious. Going back to our legal analogy, Type I errors are analogous to convicting an innocent plaintiff, while Type II errors are analogous to setting a guilty plaintiff free. Since Type I errors are more serious, it is the probability of making Type I errors that researchers evaluate when performing inferential statistics. Before performing any chemical education research, researchers should set the acceptable level (maximum probability) of making Type I errors; this is called the **level of significance** and is abbreviated α. In most chemical education research studies, the level of significance is set at .05 (i.e., the probability of finding an instructional effect that does not really exist is 1 in 20 or 5%) unless there is a reason to make this value larger or smaller. Using α values higher than .05 is very rare, and is usually reserved for exploratory studies in new areas of study where simple indications of a trend might be important (*3*).

While inferential hypothesis testing focuses on minimizing the probability of making an error (Type I or II), discussions of **power** focus on the maximizing the probability of making a correct judgment. Assuming that the effect due to an instructional method is indeed real, the power of the statistical comparison is the probability that the test will correctly reject a null hypothesis that is false. In general, the power of a statistical comparison (which is equal to $1 - \beta$) is improved by increasing the sample size (having more subjects in your study), increasing the level of significance (α, which is usually set at .05), or using directional (one-tailed) tests instead of non-directional tests (which is discouraged). The power of a test statistic is also greater when the study has a greater **effect size** (also called the **treatment effect**). The effect size is simply the difference between the actual test statistic (mean, standard deviation, etc.) and the value of the test statistic specified in the null hypothesis (not to be confused with the effect sizes used in meta-analysis studies, which are actually the **standardized effect size**, which is the effect size mentioned above divided by the standard deviation of the sample).

Step 4: Plan and Implement the Research Design

There are several excellent reference books dealing with experimental design and implementation in educational research (*9, 10*). This chapter focuses on a few key ideas specific to chemical education research using inferential statistics. These topics include determining the reliability and validity of test questions, random versus group assignment of students, and minimizing biases in educational research. Any chemical education researchers who are going to perform research involving the use of human subjects *must* get approval from their school's Institutional Review Board (IRB) before collecting data (*11*).

Chemical education researchers are typically asked to demonstrate that the test instruments used in their studies are valid and reliable. **Validity** is the extent to which a question measures what it purports to measure. There are several forms of validity (face validity, content validity, construct validity, etc.), and different forms become relevant for different test instruments. **Reliability**, on the other hand, is the extent to which the question is stable over time, i.e., if the instrument were administered to similar groups of students at two different times, highly reliable instruments would yield similar results. Reliability is more commonly used for a series of multiple-choice questions than for a single multiple-choice question or for open-ended free-response questions. Reliability measures of multiple-choice tests are used to ensure that students scoring well on the test understood the material and those scoring poorly did not (i.e., that their scores are not based on chance). Since students are unlikely to "guess" the right answer for open-ended free-response questions involving several sub-parts, measuring the reliability of these questions is not as important. Scantlebury and Boone's chapter on the Rasch method provides several techniques for evaluating reliability and validity (*12*).

Misconception—Students must be randomly assigned to treatment and control groups to get valid results: Random samples from the population are not the goal of educational research; *representative* samples are. Representative samples consist of a sample of students who have the probability of being similar to the population as a whole on the key variables relevant to the research hypotheses being investigated. Representative samples are not identical to the whole population or even to each other; they are just likely to be similar to the population (and each other) on the key variables of interest. Although random sampling or random assignment of students is one way of achieving representative samples, unless the samples are extremely non-representative, the results of a research study without random assignment can provide important information about the populations as a whole (*10*).

A more common method used in chemical education research is random cluster or group sampling (*10*), in which groups of students (classes) are randomly assigned to different instructional groups instead of randomly assigning individual students. This method is often quicker and easier than individual random assignment, and more realistic given the practical constraints of dealing with existing classes and getting IRB approval. However, it may be harder to demonstrate that the groups are representative of the population when using group-sampling techniques. As a result, these research studies are sometimes called 'quasi-experimental', and researchers must recognize that because the students were not randomly assigned, the two groups may be significantly different from each other or from the population as a whole. One way to mitigate possible group differences is to collect pretest data using the dependent variables in the study or other relevant variables, and use statistical methods (like *t*-tests or ANCOVAs) to correct for any initial differences.

The two alternatives of randomly assigning individual students into groups (which leads to more representative samples) or randomly assigning intact classes of students into groups (which is easier and less obtrusive to the teachers and students involved in the study) demonstrate the complexity of chemical education research in which the researcher trades internal validity for external validity. **Internal validity** is the extent to which a researcher has controlled extraneous variables in a study; **external validity** is the extent to which the findings of a particular study can be applied to other settings. The more experimental control a researcher employs (random assignment, controlling classroom setting, etc.), the more confident the researcher can be that any significant differences identified are actually due to the variable(s) of interest. However, the more experimental control a researcher employs, the less realistic the study is and the less representative it is of actual classroom situations that students and instructors normally face, and the less confident the researcher is that the results from the "controlled environment" will actually be seen in a normal, uncontrolled classroom environment. Chemical education researchers usually try to balance these two concerns by using real classroom environments where some control over extraneous variables is attempted. Another way chemical education researchers can address both forms of validity is to perform initial studies with high degrees of experimental control and then perform subsequent studies using more realistic settings.

Misconception—Researchers can eliminate biases from educational research: It is impossible to eliminate biases from chemical education research studies (or benchtop chemical research studies), and it is naïve to think that simply because a possible source of error (bias) in a study has been found, the study is inherently flawed and therefore useless. The best chemical education researchers can do is to attempt to minimize these errors or biases, and to recognize that these errors or biases may have some effect on the results and the conclusions based on these results.

An example from the chemical education literature illustrates this idea. Kelly, Phelps, and Sanger (*13*) used the can-crushing animation created for the study described above (*6*) to test how viewing the animation would affect high school students' macroscopic, symbolic, and microscopic understanding of the demonstration. Nine classes of students at the same high school were split into two groups. Four of these classes (taught by the same instructor) did not view the animation, while the other five (taught by a different instructor) did. What are the possible sources of error (biases) in this study design? Since the researchers used existing class sections instead of randomly assigning students or the intact class sections to the two groups, there could be an issue of non-equivalence of the two groups. Also, there could be an instructor effect since the students in the two groups received instruction from different instructors.

Why did the researchers design the study the way they did? This design was used because the researchers believed that having each teacher deliver only one type of lesson would provide maximum consistency (constancy) of instruction. They also decided to have the previously assigned instructors teach their own classes to minimize the issues associated with students receiving instruction from an unfamiliar teacher.

How did the researchers attempt to minimize the biases in this study? To address the issue of non-equivalence of the groups, the researchers collected pretest data on the students' general knowledge and ability using the Iowa Test of Educational Development. A comparison of these scores showed no statistical difference. Also, the nine classes were selected from the same high school during the same semester to minimize population differences in the two groups. To address the issue of instructor effects, the teachers planned how they would perform the can-crushing demonstration and how they would explain it at the molecular level to ensure that the two lessons were as similar as possible (with the exception of the animation). Although there were five chemistry teachers at this high school, these two teachers were chosen because they were very similar to each other in training, experience, and pedagogical beliefs. For example, both had been teaching for about five years, both graduated from the same teacher education program at the same university, both had taken several chemistry content and educational methods courses from the same instructors, and both believed in the importance of incorporating inquiry-based lessons as part of their everyday instruction.

How could the researchers have planned this research study differently? The researchers could have randomly assigned students to the two groups, or could have randomly assigned the nine classes to the two groups. Assigning students to the two groups would have been very time- and labor-intensive, and may not have been feasible in this school setting. Since these classes were taught at different times, random assignment of students would have required two instructors and two classrooms available for each class, which was not possible. But what about randomly assigning the nine classes to the two groups? This could have been designed in one of two ways: (1) Each class could have received instruction (animation or no animation) from their existing teacher and both teachers would have taught both lessons; (2) One instructor could have taught the animation lessons and the other instructor would have taught the non-animation lessons, and students may or may not have received instruction from their existing teacher. A different teacher could have been brought in to teach all of the lessons (or one teacher for the animation lessons and another teacher for the non-animation lessons), but this would have introduced more variables into the study, some of which would have been hard to control.

Would these different research designs have eliminated biases from this study? While the alternative research designs described above may have minimized some of the biases in that study, these designs have biases of their own which are neither better nor worse than those in the actual study, simply different. Design (1) requires both teachers to teach both lessons. Did both teachers teach animation lessons similarly? Did they both teach the non-animation lessons similarly? When the same teacher teaches both lessons, this introduces a whole new set of biases (*10, 14*). What if one teacher believed that the animation was better than the non-animation lesson? It is possible that teacher would (subconsciously or not) teach the animation lesson better or more enthusiastically than the other lesson. If the teacher told the students that the animations should make them learn better, the students might be more on-task, which could affect their learning (the *Pygmalion Effect*). Simply by using a new instructional method like animations, students could recognize that they are being studied and focus their attentions on the new lesson (the *Hawthorne Effect*). If one teacher disliked the animations, he or she might (subconsciously or not) try to teach the other lesson better or more enthusiastically than the animation lesson (the *John Henry Effect*).

Because Design (1) requires both teachers to teach both lessons, the researchers would need to assure that both teachers are teaching each lesson in the same manner, and now there would be two sets of instructor biases to address. Design (2) would minimize the issues associated with the same teachers teaching both lessons (i.e., the *intra*-teacher effects are gone, but now there could be *inter*-teacher effects). And now there is the issue that some students would receive instruction from a new instructor (which could lead to the Hawthorne Effect), while others would learn from a familiar instructor. The whole point of this discussion is to show that it is virtually impossible to eliminate instructional biases (whether teacher-based or not) in any chemical education study. Therefore, researchers must do what they can to identify these biases, minimize them to the greatest extent possible, and recognize how these possible biases might affect the validity of the research results. This is especially important if the chemical education research study was done in a naturalistic setting, in which the real-world constraints of the classroom setting cannot be removed in favor of more rigorous and controlled situations.

Step 5: Choose an Appropriate Test Statistic

Misconception—Only one statistical test can be valid or meaningful for any research question: This idea usually appears as the belief that whenever comparing two groups of student scores, a particular test statistic must be used (i.e., t-tests must be used when comparing two groups of students, ANOVAs must be used for more three or more groups of students, etc.). For any particular research question, several statistical tests could give you valid information. Many times, the way the data was collected (test scores, student frequencies, etc.) dictates which test statistics would be most appropriate for the data analysis. The belief that only one statistic can ever be appropriate for a given set of data also fails to recognize that many test statistics are equivalent to each other under certain conditions. For example, when performing t-tests with a large sample of students (more than 120), the test statistic is identical to the z-test statistic (2). Similarly, when comparing two independent sets of students, the independent measures t-test and the one-way ANOVA (an F-test) are equivalent, and related by the formula $t^2 = F$ (3). There are other mathematical formulas used to convert from one test statistic to another: Fisher's z-transformation converts a correlation coefficient (r) to a z-score or t-score (*2*), and there is a formula to convert the chi-square (χ^2) test statistic to z-score (*2*).

<u>*Misconception—Performing several simple statistical comparisons is just as reliable as performing one more complex comparison:*</u> This is usually seen when chemical education researchers are trying to compare three or more groups of students. One way to do this would be to compare each set of students together using a series of t-tests; another way to do this would be to compare all of the groups at once using an analysis of variance (ANOVA). Unfortunately, these two methods are not equally reliable; in particular, the first method leads to a greater probability of rejecting a null hypothesis that was actually true (Type I error). This is because the probability of making a Type I error among several comparisons is cumulative throughout a research experiment. Therefore, chemical education researchers should use a single test to make multiple comparisons instead of performing several independent tests whenever possible.

The goal of this section is to introduce common statistic tests used in chemical education research (correlation coefficients, t-tests, ANOVAs and ANCOVAs, tests of proportions, and some non-parametric tests), and describe specific examples of studies from the chemical education literature that have used these tests. The list of statistics tests described here is not intended to be exhaustive; it simply reflects the statistical tests most commonly used by chemical education researchers.

Correlation Coefficients

Correlation coefficients are used to look for relationships between two variables, and the most common correlation coefficient used is the Pearson product-moment correlation coefficient (r). When calculating correlation coefficients, the two variables must be at the interval or ratio level (*2*), which means that correlation coefficients cannot be used with category data that are dichotomous (mutually exclusive) and non-numerical (like animation/non-animation group, male/female, single/married/divorced, etc.). Values for the Pearson r vary from -1 to $+1$. Negative r-values imply negative correlations (as one variable increases, the other decreases) while positive r-values imply positive correlations (as one variable increases, so does the other and vice versa); r-values of 0 imply no relationship between the two variables. It is important to note that Pearson r-values assume *linear* relationships between the two variables; if non-linear relationships are expected or observed, correlation ratios (η) that recognize non-linear relationships can be calculated (*10*).

Williamson and Abraham (*15*) investigated the relationship between introductory college chemistry students' scores on the Test of Logical Thinking,

TOLT (which measures students' abilities to control variables and think using proportions, combinations, probabilities, and correlations) and their scores on two sub-categories of the Particulate Nature of Matter Evaluation Test, PNMET, (which requires students to make drawings, give explanations, and choose from multiple choices explaining chemical phenomena). They found that there was a significant positive correlation between students' TOLT scores and their scores on the PNMET Solids, Liquids, and Gases sub-section ($r = 0.44$) and the Reaction Chemistry sub-section ($r = 0.52$). Tasker and Dalton (*16*) showed that visuospatial working-memory capacity (measured by the Figural Intersection Test) and students' post-test knowledge (measured by a post-test created by the researchers) were positively related to each other ($r = 0.59, p = .05$).

Misconception—A statistically significant correlation means one variable is causing the changes in the other variable: A common mistake made by chemical education researchers using correlation coefficients is to assume that simply because there are statistically significant relationships between two variables that this means that changes in one variable "caused" the changes in the other variable (i.e., a cause-and-effect, or *causal*, relationship). While there could be casual relationships between these variables, they are not guaranteed. These relationships could be coincidence, or they could be the result of another unexamined variable. An example of coincidental correlations caused by an unrelated variable involves a study in which a researcher tries to measure the relationship between wearing blue shirts when taking a test and failing the test. After collecting data in several classes of different sizes taught by the same teacher, the researcher finds a statistically significant positive correlation: In classes with more students wearing blue shirts, more students fail the test. Does this mean that wearing a blue shirt caused the students to fail? Does it mean that failing the test somehow turned some students' shirts blue? The answer to both of these questions is no. The flaw in this research is that the researcher neglected to control other variables (like the total number of students in the class), and although a statistical correlation between shirt color and failure rate exists, the real cause of the relationship is due to changes in class size.

Student *t*-Tests

When comparing mean scores of one group of students to specific values (one-sample case) or to mean scores of another group of students (two-sample case), chemical education researchers can calculate z-scores only if the standard deviation of the population as a whole is known. Since population standard deviations are almost never known, researchers must estimate these values using sample standard deviations. When this is done, researchers calculate t-scores instead of z-scores. While the z-distribution (*normal* distribution) is independent of the number of subjects in the study, t-distributions change based on the numbers of subjects in the study. The effect of the sample size on statistical comparisons is usually described in terms of the **degrees of freedom** (df). The degrees of freedom are a measure of the number of scores in the sample that are allowed to vary. In the case of t-tests, df are simply $N-1$ (where N is the total number of subjects in the study) for a one-sample case, and $N-2$ for a two-sample case.

One-sample t-tests are used by researchers to compare mean scores of one sample of students to specific numbers. For example, one-sample t-tests allow teachers to determine whether a class performed differently on a standardized ACS chemistry test than the reported mean. In **repeated (dependent) measures** t-tests, a group of students provide two different responses, usually before and after a treatment has been administered to the class. Incomplete data are usually discarded. Difference (or delta) scores are calculated as post-score minus pre-score, and the difference scores are compared to zero. Mean difference scores that are statistically different from zero imply a difference in the two scores—positive values mean the treatment improved student responses, while negative values mean the treatment lowered student scores. When researchers want to compare the means of two different groups of students, **independent measures** t-tests are used. These tests can be used to compare pre-test scores of two different groups of students before an instructional treatment has been administered. If the groups are not found to be different, then there is no need to correct for initial differences; if they are found to be different, researchers need to correct for these initial differences. A combination of repeated measure and independent measures t-tests can be used in which difference (delta) scores of two different groups are compared using an independent measures t-test. This design is popular because it allows the comparison of two groups of students but also corrects for any initial differences.

Every statistical test has some assumptions regarding the data and conditions under which they were collected. For t-tests, it is assumed that the sample has been randomly selected from the population (so that the sample is representative of the population), and that the population is normally distributed. The independent measures t-test also assumes that the two populations from which the samples have been selected have the same variance and standard deviation. Thankfully, t-tests are considered to be fairly *robust* as long as each sample is relatively large ($N > 30$), which simply means that they still provide valid statistical measures even when there are departures from the assumptions.

Russell et al. (*17*) used a repeated measures t-test to determine the effectiveness of animations showing synchronous depictions of chemical reactions using the macroscopic, microscopic, and symbolic representations on students' conceptual chemistry knowledge. Students in two sections of college introductory chemistry were given a pre-test, then received instruction using these animations, and answered similar questions on a post-test. A comparison of pre- and post-test scores (presumably done using difference scores) showed a significant difference between the pre-test and post-test scores ($t_{294} = 15.61, p < .0001$). The subscripted number is the *df* in this study; the positive *t*-score showed that the animations increased the students' scores. Sanger and Badger (*18*) used several independent *t*-tests to investigate the effectiveness of 3-D electron density plots and electrostatic potential maps on students' conceptual understanding of molecular polarity and miscibility. One group received instruction using these visualization techniques while another group received instruction without these techniques. The authors used students' scores on the 1997 ACS Special Exam (2nd term) as a pre-test, and compared them using an independent measures *t*-test. The results ($t_{65} = 0.07, p = .94$) suggested that the two groups did not have significantly different chemistry content knowledge. Because the groups were not significantly different before the instruction, the authors did not need to correct for initial student differences. Comparison of the responses from the two sets of students showed that students who viewed the electron density plots and electrostatic potential maps provided more correct descriptions of particle attractions in aqueous sodium chloride ($t_{70} = 6.80, p < .0001$) and in a mixture of soap, water, and grease ($t_{66} = 2.67, p = .0048$).

Analysis of Variance and Analysis of Covariance

While *t*-tests allow chemical education researchers to compare mean scores from one or two groups of students, sometimes they are interested in comparing more groups. Researchers could perform several *t*-tests comparing two groups at a time, but that would increase the possibility of making a Type I error. To prevent this, researchers should use **analysis of variance** (ANOVA) tests, which allow researchers to compare all groups at one time with one statistical test. When using two sets of independent students, independent measures *t*-tests and one-way ANOVAs give the same information and are related ($t^2 = F$). The null hypotheses for ANOVAs assume that the mean scores for all groups are the same; the research hypotheses state that they are different but do not specify *how* they are different. When *F*-statistics are calculated using ANOVAs, there are *two* different degrees of freedom that must be reported: The first is called the **between-treatments degrees of freedom** ($df_{between}$) and equals $k - 1$, where k is the total number of categories within the independent variable; the second is called the **within-treatments degrees of freedom** (df_{within}) and equals $N - k$, where N is the total number of subjects. Both of these *df* values must be known in order to evaluate the significance of a particular *F*-value.

For **one-way ANOVAs**, chemical education researchers can compare mean scores for several groups of students ($n \geq 2$) based on differences in one independent variable. **Repeated (dependent) measures ANOVAs** are used when data for the different treatment methods were collected from the same group of students (similar to repeated measures *t*-tests). Repeated measures

ANOVAs calculate the same F-values as independent ANOVAs, but also calculate an F-value for the subjects. The null hypotheses for these statistics are that there is no relationship between the subjects and the dependent variable. Because repeated measures ANOVAs correct for the effect of the subjects, the df_{within} value for repeated measures ANOVAs are decreased from $k(n-1)$ for independent ANOVAs to $(k-1) \times (n-1)$, where k is the number of responses from each subject, and n is the total number of subjects. The "lost" degrees of freedom are now used to describe the subjects: $df_{subjects} = n - 1$.

ANOVAs have one more advantage over t-tests: ANOVAs can compare mean scores of several groups of students based on differences in more than one independent variable. When two independent variables are studied, the test statistics are referred to as **two-way ANOVAs** (higher-order ANOVAs are possible, but are rarely used by chemical education researchers). The major advantage of performing two-way ANOVAs, instead of two separate one-way ANOVAs or t-tests, is that two-way ANOVAs can determine whether there is a difference due to each of the independent variables (called a *main effect*) and whether there is an interaction between the two independent variables. This occurs when the effects of one of the variables depends on the other variable (e.g., the effect of an instructional lesson may be different for males and females). The null hypothesis is that there is no interaction between variables, and the research hypothesis is that there is some sort of interaction. For two-way ANOVAs, there are three F-values calculated: The main effect for variable A (F_A), the main effect for variable B (F_B), and the interaction between A and B ($F_{A \times B}$). The df_{within} value is the same for the three tests: $df_{within} = N - k_A \times k_B$, where N is the total number of subjects, k_A is the total number of categories within variable A, and k_B is the total number of categories within variable B. The $df_{between}$ values are $(k_A - 1)$ for the main effect of variable A, $(k_B - 1)$ for the main effect of variable B, and $k_A \times k_B$ for the interaction between A and B.

Analysis of covariance tests (ANCOVA) can be used to correct for any initial differences among the groups using pre-test data. Chemical education researchers attempt to minimize initial differences by adequate experimental research design; however, sometimes it is difficult to have sufficient control of the groups (especially existing classes) or the researchers may have reason to suspect that, despite the research design, these groups are different in some important way. ANCOVAs allow researchers to correct for initial differences by statistical methods. ANCOVAs are simply ANOVAs that include an additional independent variable called the **covariate**. ANCOVAs assume that the effect of the covariate on the dependent variable is a linear relationship and that the covariate is unaffected by the other independent variables. ANCOVAs calculate the same F-values as ANOVAs, but they also calculate an F-value for the covariate. The null hypothesis for the covariate assumes no relationship between the covariate and the dependent variable. Because ANCOVAs correct for the effect of the covariate, df_{within} is decreased by one compared to the corresponding ANOVA and $df_{covariate} = 1$ (the $df_{between}$ values remain the same).

If a significant main effect has been identified for ANOVAs or ANCOVAs, chemical education researchers can be confident there is a statistical difference in the scores based on that independent variable. But *how* are these mean scores different? To answer this question, researchers perform **post-hoc comparisons**. Research questions asked before the experiment are called *a priori* questions; tests planned after the experiment (usually during data analysis) are called post-hoc comparisons (or *a posteriori* questions). It is appropriate to make post-hoc comparisons *only* when ANOVAs or ANCOVAs have identified a significant main effect. If there is no main-effect, then the post-hoc comparisons will be meaningless—it doesn't make sense to ask how the subgroups are different if you did not conclude that they *are* different. The most common post-hoc comparison tests (*2*) are the Tukey method (which requires an equal number of subjects in each subgroup) and the Tukey/Kramer method (for unequal numbers of subjects in each subgroup). Both methods make several pair-wise comparisons of each subgroup to determine if their mean scores are significantly different. Unlike pair-wise comparisons using *t*-tests, these tests do not increase the likelihood of making a Type I error.

Unfortunately, there are no post-hoc comparison tests to explain significant interaction effects identified for two-way ANOVAs or ANCOVAs. In this case, chemical education researchers usually create an **interaction plot** (Figure 1). In interaction plots, the mean score for each subgroup is placed on the vertical axis, and one of the independent variables is placed on the horizontal axis at equal intervals. Data points from each subgroup of the other independent variable are connected with lines. If there were no interaction between the two variables, these lines would be parallel. These plots provide the researcher with a visual means of identifying the effects of one independent variable on the other.

The assumptions made regarding data collection for ANOVAs and ANCOVAs are similar to those for *t*-tests—the samples have been randomly selected from the population, the population and populations of the samples are normally distributed, and the variances and standard deviations of the subgroups are homogeneous. As with *t*-tests, ANOVAs and ANCOVAs are relatively robust with respect to non-normality and are relatively robust to heterogeneity of variances when the sample sizes are the same within each subgroup (*19*).

Williamson and Abraham (*15*) studied how the use of computer animations affected students' conceptual understanding of chemistry. This study used three groups of students: One group did not view animations (control), one group viewed animations in lecture only (lecture), and one group viewed them in lecture and recitation (lecture/recitation). All students answered two subsections of the PNMET, a test measuring students' particulate-level chemistry understanding. Because the authors were concerned that students' abilities to think at the formal operational level (*20*) might affect their answers, they used

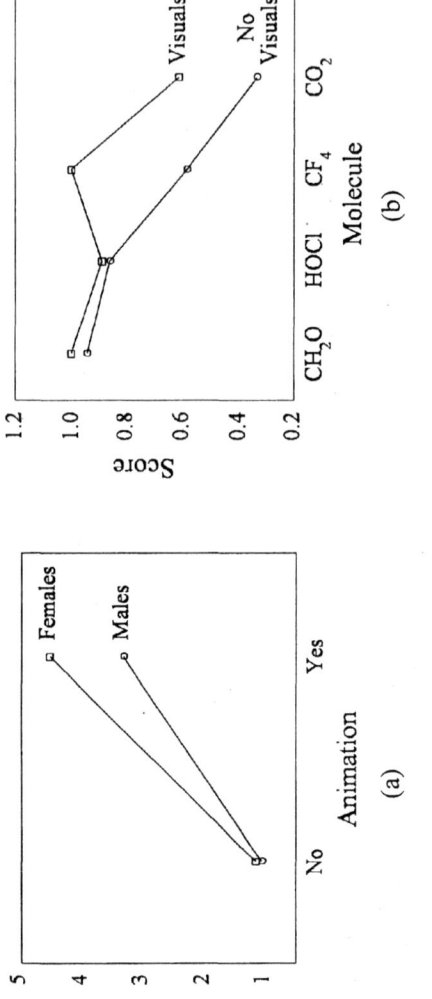

Figure 1. Interaction plots for significant interactions: (a) Interaction between animation treatment and gender (Reproduced from reference 21. Copyright 2006 ACS Division of Chemical Education, Inc.). (b) Interaction between molecule type and visual treatment (These data are from reference 18. Copyright 2001 ACS Division of Chemical Education, Inc.)

the TOLT as a covariate. They performed one-way ANCOVAs for the two subsection scores of the PNMET, using group as the dependent variable and TOLT scores as the covariate ($N = 124$). For both subsections, the covariate was significantly correlated to student performance ($F(1,120) = 28.31$, $p < .0001$; $F(1,120) = 41.68$, $p < .0001$). Students in the three groups performed differently on both subsections ($F(2,120) = 4.57$, $p = .012$; $F(2,120) = 4.21$, $p = .017$). Post-hoc tests using a Games-Howell statistic showed that the control group had significantly lower mean scores than both the lecture and lecture/recitation groups for both subsections, and the differences between the lecture group and lecture/recitation group were not significantly different for either subsection.

Yezierski and Birk (21) tested the effect of gender and the use of computer animations on students' scores on the Particulate Nature of Matter Assessment (ParNoMA), an instrument testing students' molecular-level understanding of chemistry. The subjects ranged from eighth-graders to first-year college students. The treatment group received instruction including four animations of water in various states of matter; the control group received equivalent instruction without computer animations. The authors used a one-way ANOVA on pre-test scores to find that there were no initial differences between the two groups ($F(1,717) = 2.01$, $p = .16$). Because there were no significant initial differences, the authors used an ANOVA instead of an ANCOVA. They compared the mean difference (delta) scores for the ParNoMA using a two-way ANOVA with the difference scores as the dependent variable, treatment group as one dependent variable, and gender as the other ($N = 719$). This ANOVA is called a 2 × 2 ANOVA; the numbers represent the number of subgroups in each independent variable (animation/no animation and male/female). The researchers found a significant main effect for treatment group ($F(1,715) = 118.07$, $p < .001$), a significant main effect for gender ($F(1,715) = 6.38$, $p = .012$) and a significant interaction of treatment and gender ($F(2,715) = 4.64$, $p = .032$). Because the treatment and gender variables have only two subgroups, there is no need to perform post-hoc tests; the ANOVA showed that these groups are significantly different from each other. Students who viewed the animations performed better than those who did not (3.98 versus 1.12, respectively), and females showed greater improvement than males (2.91 versus 2.15, respectively). The interaction plot for treatment and gender appears in Figure 1a. The mean scores for males and females who did not view the animations are similar, but for students who viewed the animations females improved much more than males. These results suggest that animations could be used to shrink the existing gender achievement gap in chemistry.

Sanger and Badger (18) used a repeated-measures two-way ANOVA to determine the effect of visualization strategies on students' responses to four questions. One group of students viewed several 3-D electron density plots and electrostatic potential maps to help them visualize charges on individual atoms within a molecule; the other group received similar instruction without these visualization techniques. Students were asked to categorize four molecules (CH_2O, $HOCl$, CF_4, and CO_2) as either polar or non-polar. Students who saw the visuals scored significantly higher than students who did not ($F(1,52) = 10.57$, $p = .002$). Students' responses in the two groups were significantly different for the four molecules ($F(3, 156) = 23.16$, $p < .0001$), and there was a significant interaction between the visuals and the four questions ($F(3,58) = 3.21$, $p = .025$). The interaction plot (Figure 1b) shows that the scores for polar molecules (CH_2O and $HOCl$) were similar for both sets of students; however, students who viewed the visuals were better at identifying symmetrical molecules with polar bonds as being non-polar (CF_4 and CO_2) than students who did not view the visuals.

Tests of Proportions

Tests of proportions are statistical tests used to compare the fraction (proportion, percentage) of students within a sample providing a certain answer. This test requires dichotomous, mutually-exclusive independent variables so all subjects fall into only one of the two categories. Examples include gender, race (if categorized as white/not white, etc.), placement in one of two instructional groups, response to question (right/wrong, true/false), etc. These tests use the binomial distribution, but if the sample size is reasonably large the normal distribution is an adequate approximation of the binomial distribution. Just as with t-tests, chemical education researchers can compare the proportion of a sample to a specific number (**one sample test of proportions**), to the proportion of another independent sample of students (**independent test of proportions**), or another proportional response from the same sample of students (**repeated, or dependent, test of proportions**).

Tests of proportions are commonly used in survey research, especially by polling agencies and political pollsters trying to predict outcomes of a political election. However, they are also used in educational research to compare the proportion of students answering dichotomous questions correctly, including true-false and multiple-choice questions (categorized as right/wrong). Another appropriate way to compare proportions is to calculate a t-test based on students' responses to question. Students are given a score of 1 for the correct answer and a score of 0 for wrong answers. The resulting test statistics (z for tests of proportions, t for t-tests) are similar but not the same as each other, due in part to the differences in the binomial, normal, and t-distributions. There is at least one advantage of using tests of proportions over t-tests. If the proportion of any subject sample is exactly 0.00 or 1.00, then the standard deviation of these responses is zero (i.e., there is no variability in their answers). Standard deviations of zero make it impossible to calculate t-scores; however, test statistics for tests of proportions can still be calculated (22).

Sanger, Phelps, and Fienhold (*6*) used independent tests of proportions to analyze the differences in responses from students who viewed a computer animation of a can-crushing demonstration and from students who did not. They found that students who viewed the animation were more likely to provide correct explanations (34% versus 17%; $z = 3.02$, $p = .0025$), were more likely to recognize that the condensation of water was important (58% versus 24%; $z = 4.25$, $p < .0001$), and less likely to blindly quote ideal gas laws in their predictions (6% versus 33%; $z = -4.38$, $p < .0001$). The first two z-values were positive because the animation had a positive effect on the students' responses; the last z-value was negative because the animation had a negative effect on the proportion of students blindly quoting gas laws.

Chi-square Tests for Nominal Data

Just like tests of proportions, chi-square tests are used when the data are at the frequency or category level (numbers of subjects in mutually-exclusive categories). Unlike tests of proportions, each variable can have more than two categories, which is useful for comparing student responses to multiple-choice questions or Likert scales. Chi-square (χ^2) tests are **non-parametric**, which simply means that they do not make the parametric assumptions that the data have normal distributions, that the data are linear, or that the variances among groups are homogeneous. As with t-tests and tests of proportions, χ^2 tests can compare subjects in a particular sample to specific values (chi-square tests of goodness-of-fit), to subjects in one or more independent samples (chi-square test of homogeneity, which is also called a chi-square test of independence), or to another frequency from the same sample (McNemar's test for significance of change). All χ^2 tests involve comparing the observed frequencies for each group with the expected frequencies if the null hypothesis were true.

The null hypotheses for chi-square tests of goodness-of-fit assume that the frequency of subjects in each subgroup is the same as the proposed distribution. The research hypothesis states that this distribution is different than the proposed distribution, but does not specify *how* it is different so post-hoc test are used to explain these differences. The expected frequencies are calculated using the total number of subjects and the proposed proportions in each category. For chi-square goodness-of-fit tests, the degrees of freedom are equal to the number of categories being compared minus one: $df = C - 1$, where C is the number of categories within the independent variable. As with t-distributions (and unlike z-distributions), there are many χ^2 distributions based on the degrees of freedom. If a distribution is shown to be different than the predicted distribution (and the null hypothesis is rejected), chemical education researchers can calculate **standardized residuals** (R) to determine how the distribution is different from the proposed one. If the standardized residual for a particular category is larger than +2, the frequency of subjects in that category is larger than predicted, while a standardized residual smaller than −2 means that the frequency of subjects in that category is smaller than predicted (standardized residuals between −2 and +2 are not significantly different from the predicted distribution).

Chi-square tests of homogeneity (also called **chi-square tests of independence**) are used when the data include two independent variables with two or more mutually-exclusive categories. The null hypotheses for these comparisons are that the frequency distributions of the subjects within one independent variable are *independent* of the other variable, or that the distributions within each category for the two independent variables are *homogeneous* (hence the alternative names for this test). The frequency data are placed in a **contingency table** that has the categories of one of the independent variables as columns, and categories of the other independent variable as rows. Each cell contains the frequency (number) of subjects belonging to those two categories. Calculating the expected frequencies is a little more complicated than for the one-case sample, and uses the total number of students and the sum of frequencies for each individual row and column. The degrees of freedom for chi-square tests of homogeneity is $df = (R - 1)(C - 1)$, where R is the number of categories in the independent variable in the rows and C is the number of categories in the independent variable in the columns of the contingency table. If the distributions are shown to be different from one another, standardized residuals can be calculated to explain these differences.

When chemical education researchers compare frequency data for two responses from the same students (usually before and after an instructional treatment), **McNemar's tests for significance of change** are used. They are referred to as tests for 'significance of change' because they ignore responses from subjects who have not changed their answers. For subjects who changed their answers, the null hypothesis assumes that the changes from one answer to another are randomly distributed (i.e., the probability of changing from any one answer to another is the same). Unlike independent χ^2 tests, standard residuals cannot be calculated to explain these differences, and chemical education researchers usually look at the data to make qualitative comparisons.

Comparisons of three or more mutually-exclusive category variables can be made using **multiway frequency analysis** tests (*23*). Multiway frequency analyses with two independent variables are the same as chi-square tests of homogeneity. When three independent variables are compared, multiway frequency analyses look for one-way associations (χ^2 tests of goodness-of-fit), two-way associations (χ^2 tests of homogeneity, or main effects for ANOVAs and ANCOVAs) and three-way associations (similar to interaction effects for ANOVAs and ANCOVAs). When comparing four independent variables, multiway frequency analyses look for one-, two-, three-, and four-way interactions, and so on. The loglinear model of multiway frequency analyses starts with all possible one-, two-, three-, and higher-way associations and eliminates as many of them as possible while still maintaining an adequate fit between expected and observed cell frequencies.

The study performed by Sanger et al. (24) demonstrates the different uses of chi-square tests. The data, which appear in Table I, show the frequency of student responses to a particulate gas-law question that has appeared in several chemical education research studies (24-27). Sanger et al. investigated whether viewing an animated version of the multiple-choice question would change the distribution of student responses. If the researchers were only interested in the proportion of students getting the question right, they could have used a test of proportions, but since they were interested in the distribution of all four choices, a χ^2 test was used.

The researchers needed to determine whether the distribution of student responses before seeing the animation (row 4 in Table I) was different from the previous responses (rows 1-3 in Table I). One way to make this comparison would be to use a χ^2 test of goodness-of-fit with the average of the preexisting data as the expected values. Using this comparison, the data in row 4 was not shown to be different from the expected values ($\chi^2(3) = 1.90, p = .59$). Another way to answer this research question is to perform a χ^2 test of homogeneity on the first four rows of Table I. All four data sets are compared to each other (i.e., none of the groups are used to establish expected values, and any frequency could be found to be different than the other values). The test of homogeneity showed that these four groups are statistically different from each other ($\chi^2(3) = 27.53, p = .0011$). Analysis of the standardized residuals, however, showed that none of the frequencies for fourth row were significantly different from their expected frequencies (all R values between -2 and $+2$).

Sanger et al. (24) also asked whether the distribution of student responses after viewing the animated question was different from the same students' responses before viewing the animation. An independent chi-square test could not be used to compare the data in rows 4 and 5 of Table I because they came from the same students. Therefore, the researchers used the McNemar test for significance of change. These data appear in Table II. The distribution of responses was found to be different before and after viewing the animated question ($\chi^2(3) = 51.1, p < .0001$). The data showed that while 62 of the 210 subjects (30%) changed from an incorrect option to the correct answer, only 4 subjects (2%) changed from the correct answer to an incorrect option, demonstrating that the animation had a positive effect on the distribution of student responses. Since the student responses before viewing the animation were not different from the previously reported data, the researchers could have performed a χ^2 test of goodness-of-fit for the data in row 5 of Table I using the data in rows 1-3 as the expected values ($\chi^2(3) = 52.78, p < .0001$) or a χ^2 test of homogeneity comparing the rows 1-3 and 5 ($\chi^2(9) = 67.60, p < .0001$). The data from row 4 must be left out of this comparison because it is not independent of the data in row 5. Analysis of the standardized residuals showed that two

Table I. Frequency (Percentage) of Student Responses to the Multiple-Choice Question

The diagram to the right represents a cross-sectional area of a steel tank filled with hydrogen gas at 20°C and 3 atm pressure. (The dots represent the distribution of H_2 molecules).

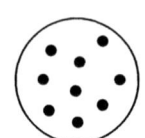

Which of the following diagrams illustrate the distribution of H_2 molecules in the steel tank if the temperature is lowered to −20°C?

Source of data	(a)	(b)	(c)	(d)
Reference 25, $N = 198$	72(36)	56(28)	50(25)	11(6)
Reference 26, $N = 285$[a]	89(31)	136(48)	34(12)	23(8)
Reference 27, $N = 330$	110(33)	141(43)	52(16)	27(8)
Reference 24, static question, $N = 210$	62(30)	91(43)	40(19)	17(8)
Reference 24, animated quest., $N = 210$	120(57)	51(24)	30(14)	9(4)

NOTE: The sum of the four choices and the total number of students are not the same for some entries; presumably, this is because some students did not answer the question.

SOURCE: Reproduced from Reference 24. Copyright 2006 ACS Division of Chemical Education, Inc.

Table II. Number of Student Responses to the Multiple-Choice Question Before and After Seeing the Animated Question

		Before Animation			
		Correct (a)	Incorrect (b)	Incorrect (c)	Incorrect (d)
After Animation	Correct (a)	58	42	10	10
	Incorrect (b)	3	41	6	1
	Incorrect (c)	1	4	24	1
	Incorrect (d)	0	4	0	5

NOTE: For statistical purposes, the data from (c) and (d) were collapsed into one group. The boxes represent the 3 × 3 grid used for the McNemar test for significance of change.
SOURCE: Reproduced from Reference 24. Copyright 2006 ACS Division of Chemical Education, Inc.

frequencies for row 5 were significantly different than their expected frequencies: The frequency in cell 5(a) (the correct answer) was significantly higher than expected and the frequency in cell 5(b) was significantly lower than expected ($R = 4.30$ and -3.22, respectively).

Kelly, Phelps, and Sanger (13) used multiway frequency analysis to determine the effect of viewing a computer animation on students' abilities to answer macroscopic, microscopic, and symbolic questions. Two groups of students received similar instruction on the properties of gases and gas laws, but one group also viewed a computer animation of the can-crushing demonstration. Students provided explanations of the demonstration, which were categorized as either 'good' or 'poor' for each chemical representation (macroscopic, microscopic, and symbolic). The researchers performed a four-way multiway frequency analysis using four independent variables (treatment group, macroscopic score, microscopic score, and symbolic score). The multiway frequency analysis showed no four- or three-way associations (all $p > .05$). Five of the six two-way associations were significant ($p \leq .05$). The animation had a significant positive effect on students' responses for all three representational levels (i.e., higher proportions of students in the animation group provided good answers to all three questions than students who did not view the animation). More students who answered the macroscopic question correctly also answered the microscopic and symbolic questions correctly. One of the one-way associations was significant, and it showed there were more students who viewed the animation than those who didn't, which was irrelevant to the study.

Since chi-square tests are non-parametric, they do not make the parametric assumptions that the data are normally distributed, that the data are linear, or that their variances are homogeneous. However, chi-square tests do make some assumptions. Chi-square tests assume that samples were randomly sampled from the population as a whole, and they require that the categories in each independent variable are mutually exclusive, so that a subject belongs in one and only one cell of the contingency table. An assumption made for χ^2 tests of homogeneity is that the samples being compared are independent of each other. A major issue that plagues chi-square tests is when the expected frequencies of cells in the contingency table are less than 5 (especially for a 2 × 2 contingency table when $df = 1$). When $df = 1$ and the expected frequencies are very small, the assumption of continuity is likely to be violated. For larger contingency tables and larger df, the lack of continuity due to small expected frequencies is not as important; however, most books agree that the possible lack of continuity is not a problem unless 20% of the cells have expected frequencies less than 5 (*2, 3*). If too many cells contain expected frequencies below 5, it is usually recommended that one or more of the categories from one or both of the independent variables be collapsed into one group. An example of collapsing categories appears in Table II. The data were collapsed by combining two of the categories together ('a', 'b', and 'c or d' instead of 'a', 'b', 'c', and 'd') for both the row data and the column data because of the low expected values for the individual cells that were collapsed.

One of the major difficulties chemical education researchers (especially those new to the field of educational research) experience is trying to decide which statistical test would be appropriate for the data they have collected. Table III provides a guideline for choosing an appropriate test statistic for continuous (real) data, given the number and types of student groups involved and the number of independent variables in the research study.

Table IV provides similar information for frequency (count, or proportion) data, given the number and types of student groups, the number of independent variables, and the number of categories within each independent variable in the research study.

Step 6: Evaluate the Null Hypothesis

In order to evaluate the null hypothesis (and either reject it or fail to reject it), chemical education researchers need to determine probability values associated with the test statistics (t, F, z, χ^2, etc.). Computerized statistics programs like SPSS or SAS report the p values along with the test statistics. If researchers calculate the test statistic on their own, they will need to refer to tables of statistical data that appear in statistics books (*2, 3*). The p values associated with test statistics represent the probability that researchers would be wrong in assuming that the null hypothesis is incorrect, and the probability that researchers would be wrong in assuming there is an effect due to the instructional lesson. In order to determine whether or not to reject the null

Table III. Choosing an Appropriate Statistical Test to Compare Mean Scores for Continuous (Real) Data

Number of Student Groups	Number of Independent Variables	Appropriate Statistical Test
One group, compare to specific scores	1	One-sample t-test
Two independent groups	1	Independent measures t-test
		One-way ANOVA
	2 or more	Multi-way ANOVA
Two dependent groups	1	Repeated measures t-test
		Repeated measures ANOVA
	2 or more	Repeated measures ANOVA
Three or more independent groups	1	One-way ANOVA
	2	Multi-way ANOVA
Three or more dependent groups	1	One-way ANOVA
	2 or more	Multi-way ANOVA

NOTE: The tests for two or more dependent groups can also be used for two or more different measures (scores) for a single group of students.

NOTE: An ANCOVA test can be substituted for any ANOVA listed in this table.

hypothesis, researchers must compare the p values for the test statistics to the level of significance (the level of error the researchers were willing to accept before the study was performed). If the p values are less than the level of significance, then researchers can reject the null hypothesis and report that the data are consistent with the research hypothesis; if the p values are greater than the level of significance, then researchers fail to reject the null hypothesis and should report that the data do not support the research hypothesis. A common mistake made by chemical education researchers is to report p values without comparing them to the level of significance or ignoring the results of these comparisons because they were not the results that were expected or desired. Misconception—Statistical significances can be qualified by saying they are 'nearly significant' or 'highly significant' depending on the p values: While it may be human nature to look at p values of .06 or even .051 and say that they are 'nearly significant' or 'approaching significance' (assuming $\alpha = .05$), it ignores the fact that based on the pre-set level of acceptable error (level of significance), the collected data do not support the research hypotheses. On the other extreme, some chemical education researchers report p values of .001 and say that they are 'highly significant' or 'very significant'. This still ignores the fact that before these studies were started, researchers set a maximum level of error that would be acceptable, and the significance of these studies (which are either 'significant' or 'not significant') is decided by these values alone. An analogy

Table IV. Choosing an Appropriate Statistical Test to Compare Frequency (Count, or Proportion) Data

Number of Student Groups	Number of Independent Variables	Number of Categories in Each Variable	Appropriate Statistical Test
One group, compared to specific scores	1	2	One-sample test of proportions
			Chi-square test of goodness-of-fit
		3 or more	Chi-square test of goodness-of-fit
Two independent groups	1	2	Independent test of proportions
			Chi-square test of homogeneity
		3 or more	Chi-square test of homogeneity
	2 or more	2 or more	Multiway frequency analysis
Two dependent groups	1	2	Repeated measures test of proportions
			McNemar's test for significance of change
		3 or more	McNemar's test for significance of change

NOTE: The tests for two or more dependent groups can also be used for two or more different measures (scores) for a single group of students.

for these mistakes would be a marksmen's competition in which archers shoot arrows at a simple target with only one ring. If the arrow falls outside this ring an archer can't say "Well, that was kind of close, give me a point anyway." And if an archer shoots an arrow right in the middle of the target he or she can't say "Wow, that arrow is so close to the middle, I'm going to draw an even smaller ring and give myself more points for hitting that ring as well." Statistical significance is an all-or-nothing, black-or-white, yes-or-no endeavor, and there is no such thing as 'almost significant' or 'highly significant' when performing chemical education research.

Step 7: Report the Results of the Study

When reporting the results of statistical analyses, chemical education researchers should report the test statistics and values, the degrees of freedom (if relevant), and the probability values for the test statistics. If some of this information is missing, then it makes it difficult for the reader (or reviewer) of the research study to be certain that the researchers have properly analyzed and interpreted the statistical data. Cooper's chapter in this book (*14*) discusses the importance of ensuring that conclusions from chemical education research studies are based on (and consistent with) the reported data.

Misconception—If a statistically significant treatment effect is found, this method should be used in every classroom: A common error made by chemical education researchers is confusing statistical significance and practical significance. This confusion usually occurs when researchers assume that if statistically significant treatment effects have been found, it means that the treatments will have profound changes in student scores ,and ultimately, their grades in a class. It also appears when researchers report student scores without performing statistical analyses and note that there are large differences in student scores, assuming that these large differences must be statistically significant. **Statistical significances** are determined from the analyses of inferential statistics; if the probability values are less than the level of significance ($p < \alpha$) then the treatments are assumed to have had a significant effect on the participants in the study. **Practical significances** are harder to define and are more subjective. Deciding practical significances requires the chemical education researcher to know more about the classroom situation than just the calculated statistics. Often, practical significances are defined based on grading scales or rubrics (i.e., practical significances lead to changes in students' grades on an assignment or for the class as a whole).

Statistical significances can be affected by several parameters. One of the most obvious is the level of significance. If chemical education researchers raise the α value, then values that were not significant can become significant; if they lower the α value, then values that were significant may become non-significant.

This is why it is important to choose a level of significance *before* performing the experiment, and definitely before performing the statistical analyses. Another factor that greatly affects statistical significance is the number of subjects in the sample (N). All other things being equal, increasing the sample size results in a smaller (more precise) estimate of the standard error because this value is based on more observations. Decreasing the value of the standard error results in a larger test statistic (z, t, F, etc.) and increases the likelihood that this difference will be statistically significant ($p < \alpha$). In general, increasing the number of subjects in the sample results in smaller and smaller differences becoming statistically significant (any difference between groups can be made statistically significant with enough subjects in the sample).

While statistical significances and practical significances are related, they are not the same thing. It is possible to have practical differences that result in changes to students' grades on an assignment that are not statistically significant (this usually happens when differences due to the treatments are large but the sample sizes are small), and it is also possible to have statistical differences that do not have enough of practical significance to change students' grades (this happens when differences due to the treatments are rather small but sample sizes are large). When describing the results of chemical education research, authors should clearly make a distinction between statistical significances (based on analyses of inferential statistics) and practical significances (based on more subjective methods). When submitting articles based on quantitative chemical education research to peer-reviewed journals, discussions of practical significances are usually optional; however, statistical comparisons and discussions of statistical significances are usually considered mandatory.

Misconception—If researchers fail to reject the null hypothesis, it proves that there is no effect due to the instructional treatment: This misconception is clearly related to the previous misconception about hypothesis testing providing absolute proof, but the message bears repeating. If chemical education researchers fail to reject the null hypothesis, this does not guarantee that there is no treatment effect or that if other studies were performed that no treatment effect would be found (because a Type II error could have been made). It simply means that, based on the data from this particular study, no treatment effect was found. Similarly, if one chemical education research study showed a significant treatment effect, that does not guarantee that all research studies testing this effect will find significant differences. This is why **replication studies** (studies where research questions from previously published experiments are tried again, often with slightly different student populations or experimental conditions) are particularly important in the field of chemical education research and worthy of publication.

Misconception—It is unacceptable for chemical education researchers to ask new research questions while analyzing the data: Research questions asked during or after data collection are also referred to as **post-hoc research**

questions (or **exploratory data analysis**). These are different than post-hoc comparisons made in ANOVA and ANCOVA analyses because they do not require researchers to find statistically significant differences before proceeding. Post-hoc research questions are usually investigated by researchers after the data have been analyzed, interesting trends have been noticed, and the researchers wondered whether these trends were statistically significant. These types of comparisons (which can be qualitative or quantitative statistical comparisons) are usually considered to be perfectly valid, although there are always exceptions. The biggest concern in performing many statistical post-hoc research questions is that, by adding several new research questions, the probability of making a Type I error increases. One way to correct for this is to use more stringent (conservative) rejection regions (*2*): $\alpha(\text{post-hoc}) = \alpha/c$, where α is the original alpha-level (usually .05), and c is the number of new post-hoc research questions being tested. So, if chemical education researchers propose to test five new post-hoc research questions, then the new (more conservative) rejection level for each of these new tests would be $p < \alpha/c = .01$.

Step 8: Relate the Results to the Existing Literature

Once chemical education researchers have come to some sort of conclusion based on the data collected as part of their study, the researchers must relate these results back to the existing chemical education research literature. Without this step, the researchers and others who read and evaluate these studies are left to wonder whether any statistical differences that are found are real and meaningful or if they are simply coincidental. Researchers should not only interpret how their studies fit within the existing body of literature, but also how they contradict previous findings and possible explanations for these discrepancies. This step also provides researchers with an opportunity to discuss possible problems or limitations of their studies (based on any of the previous steps described above), and possible chemical education research studies that should be performed to address new questions resulting from these studies or to address the limitations attributed to these studies.

Summary

While most chemists respect chemical education research studies based on quantitative statistical methods more than research studies involving less mathematical (more qualitative) methodologies, few are comfortable performing or evaluating this form of chemical education research. This chapter provides several steps to perform chemical education research involving inferential statistics. These rules are not unique to chemical education research, and are equally applicable to all educational research studies involving statistics. This chapter also describes common misconceptions demonstrated by chemical education researchers who are new to statistical comparisons, and explanations of the scientifically-accepted conceptions. In addition, this chapter provides a summary of some research studies from the chemical education literature to demonstrate how chemical education researchers decide which statistical tests to use in evaluating the research questions posed in their studies.

References

1. Bretz, S. L. In *Nuts and Bolts of Chemical Education Research*; Bunce, D. M., & Cole, R., Eds.; ACS Symposium Series; American Chemical Society: Washington, DC, 2007.
2. Hinkle, D. E.; Wiersma, W.; Jurs, S. G. *Applied Statistics for the Behavioral Sciences*, 3rd ed.; Houghton Mifflin: Boston, MA, 1994.
3. Gravetter, F. J.; Wallnau, L. B. *Statistics for the Behavioral Sciences;* West Publishing: St. Paul, MN, 1985.
4. Bunce, D. M. In *Nuts and Bolts of Chemical Education Research*; Bunce, D. M., & Cole, R., Eds.; ACS Symposium Series; American Chemical Society: Washington, DC, 2007.
5. Towns, M. H. In *Nuts and Bolts of Chemical Education Research*; Bunce, D. M., & Cole, R., Eds.; ACS Symposium Series; American Chemical Society: Washington, DC, 2007.
6. Sanger, M. J.; Phelps, A. J.; Fienhold, J. *J. Chem. Educ.* **2000**, *77*, 1517-1520.
7. Abraham, M. R. In *Nuts and Bolts of Chemical Education Research*; Bunce, D. M., & Cole, R., Eds.; ACS Symposium Series; American Chemical Society: Washington, DC, 2007.
8. Williamson, V. M. In *Nuts and Bolts of Chemical Education Research*; Bunce, D. M., & Cole, R., Eds.; ACS Symposium Series; American Chemical Society: Washington, DC, 2007.
9. Kelly, A. E.; Lesh, R. A. *Handbook of Research Design in Mathematics and Science Education*; Erlbaum: Mahwah, NJ, 2000.
10. Borg, W. R.; Gall, M. D. *Educational Research: An Introduction*, 4th ed.; Longman: New York, 1983.
11. Sawrey, B. A. In *Nuts and Bolts of Chemical Education Research*; Bunce, D. M., & Cole, R., Eds.; ACS Symposium Series; American Chemical Society: Washington, DC, 2007.
12. Scantlebury, K; Boone, W. J. In *Nuts and Bolts of Chemical Education Research*; Bunce, D. M., & Cole, R., Eds.; ACS Symposium Series; American Chemical Society: Washington, DC, 2007.

13. Kelly, R. M.; Phelps, A. J.; Sanger, M. J. *Chem. Educator,* **2004**, *9*, 184-189.
14. Cooper, M. M. In *Nuts and Bolts of Chemical Education Research*; Bunce, D. M., & Cole, R., Eds.; ACS Symposium Series; American Chemical Society: Washington, DC, 2007.
15. Williamson, V. M.; Abraham, M. R. *J. Res. Sci. Teach.* **1995**, *32*, 521-534.
16. Tasker, R.; Dalton, R. *Chem. Educ. Res. Pract.* **2006**, *7*, 141-159.
17. Russell, J. W.; Kozma, R. B.; Jones, T.; Wykoff, J.; Marx, N.; Davis, J. *J. Chem. Educ.* **1997**, *74*, 330-334.
18. Sanger, M. J.; Badger, II, S. M. *J. Chem. Educ.* **2001**, *78*, 1412-1416.
19. Glass, G. V.; Peckham, P. D.; Sanders, J. R. *Rev. Educ. Res.* **1972**, *42*, 237-288.
20. Nurrenbern, S. C. *J. Chem. Educ.* **2001**, *78*, 1107-1110.
21. Yezierski, E. J.; Birk, J. P. *J. Chem. Educ.* **2006**, *83*, 954-960.
22. Sanger, M. J. *J. Chem. Educ.* in press.
23. Tabachnick, B. G.; Fidell, L. S. *Using Multivariate Analysis,* 3rd ed.; Harper Collins: New York, 1996; pp 239-319.
24. Sanger, M. J.; Campbell, E.; Felker, J.; Spencer, C. *J. Chem. Educ.* in press.
25. Nurrenbern, S. C.; Pickering, M. *J. Chem. Educ.* **1987**, *64*, 508-510.
26. Sawrey, B. A. *J. Chem. Educ.* **1990**, *67*, 253-254.
27. Sanger, M. J.; Phelps, A. J. *J. Chem. Educ.* in press.

Chapter 9

Mixed Methods Designs in Chemical Education Research

Marcy Hamby Towns

Department of Chemistry, Purdue University, 560 Oval Drive, West Lafayette, IN 47907

Mixed methods designs allow researchers to use both qualitative and quantitative methods in the same study in order to balance the inherent strengths and weaknesses of each research methodology. The sequential or concurrent engagement of both research methodologies can lead to more interpretable and valid outcomes than either approach could provide alone. Multiple forms of data and analysis require very specific research designs as well as the careful consideration of how the data and analysis will be will be combined and interpreted. This chapter combines and extends the discussion of quantitative and qualitative methodologies in the preceding chapters. Examples from chemistry are used to highlight the design decisions researchers may encounter. Mixed methods studies from the research literature are included as references for those interested in designing, presenting, and publishing this type of research.

Introduction

Mixed methods research suggests exactly what the name implies—the mixing of two methods, in this case quantitative and qualitative methods of data collection and data analysis in a single study. However, this is a deceptively simplistic definition and use of this research design encompasses unique questions that need to be addressed. Answers to such questions can lead to a clarification and understanding of 1) the rationale for using a mixed methods approach and 2) the decisions that must be made in designing such a study. Some of the questions that must be addressed are as follows: What research questions are best answered by a mixed methods design? How are methods mixed in a single study—sequentially or concurrently? Should the qualitative data be collected first, or the quantitative, when the data is collected sequentially? In what order should the data be analyzed and interpreted? How can this approach be used to expand understanding of the phenomenon under investigation? In what ways do these differing methods of data collection and analysis enable convergence and confirmation of findings?

Key Decisions In Mixed Methods Designs

When a research study is designed, the research questions dictate the data collection and analysis methodology. For example, Sanger and Badger ([1]) asked "how the use of visualization strategies associated with dynamic computer animations and electron density plots affects students' conceptual understanding of molecular polarity and miscibility." To answer this research question a quantitative methodology using a two-factor repeated measures ANOVA was used. Two groups were compared—one that received instruction using computer animations and electron density plots while the other used static drawings, wooden models, and physical demonstrations. For this study a quantitative methodology fit the research question best.

Other research questions are better suited to a design that uses both qualitative and quantitative methods. Mulford and Robinson ([2]) developed "an instrument to measure the extent of students' alternate conceptions about topics" found in a traditional first semester general chemistry course. Initially, an instrument consisting of 18 free-response questions was piloted. Qualitative analysis of the responses allowed the researchers to develop a multiple-choice survey instrument with incorrect answers that reflected the conceptions of students. The survey was used in a pre-test and post-test format with general chemistry students and the results were analyzed quantitatively to document changes in alternate conceptions among the students. This study was well matched to a mixed methods design where qualitative methods were used to develop an instrument and quantitative methods were used to interpret data acquired from the instrument.

In each of these studies key design decisions were made with regard to which methodology or methodologies best fit the research question. In the field of mixed methods research Creswell (*3*) identifies four key decisions, as illustrated in Figure 1, that researchers are required to make:

- How will data collection be implemented?
- Which research approach has the dominant priority?
- How will data collection and analysis be integrated?
- What theoretical framework will guide the study?

Implementation

In the *implementation* step, researchers must decide whether the data is to be collected sequentially or concurrently. There should be a clear rationale for choosing a specific strategy that is tied to the overall goal of the study. For example, in a sequential design where the qualitative data are collected and analyzed first, the emergent understandings may be explored with a wider audience in a second quantitative phase. That was the implementation approach used in the Mulford and Robinson study (*2*) described above. The qualitative phase took place first and was used to develop the survey implemented in the quantitative phase of the study.

In a concurrent study, the qualitative and quantitative data collection and analysis take place simultaneously with the outcomes continuously informing each other. Bunce, VandenPlas, and Havanki (*4*) used this design when investigating the impact of a student response system and WebCT quizzes on student achievement and attitudes. A variety of quantitative measures were augmented with free response questions in the data collection phase. The combination and integration of quantitative and qualitative findings expanded the breadth and depth of the outcomes from the study.

Priority

An important aspect of mixed method designs is the *priority* of the quantitative and qualitative approach. In other words, does one research approach have a dominant priority over the other or are they of equal priority? The emphasis of either approach is dictated by the intent of the researcher and the goals of the study. In a practical sense, the first type of data collection usually has the dominant priority. For example, in the Mulford and Robinson

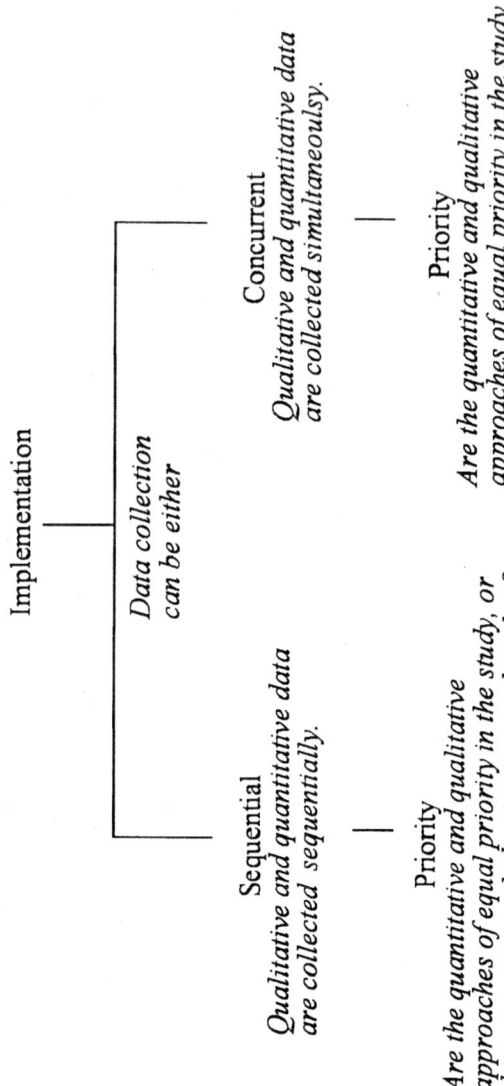

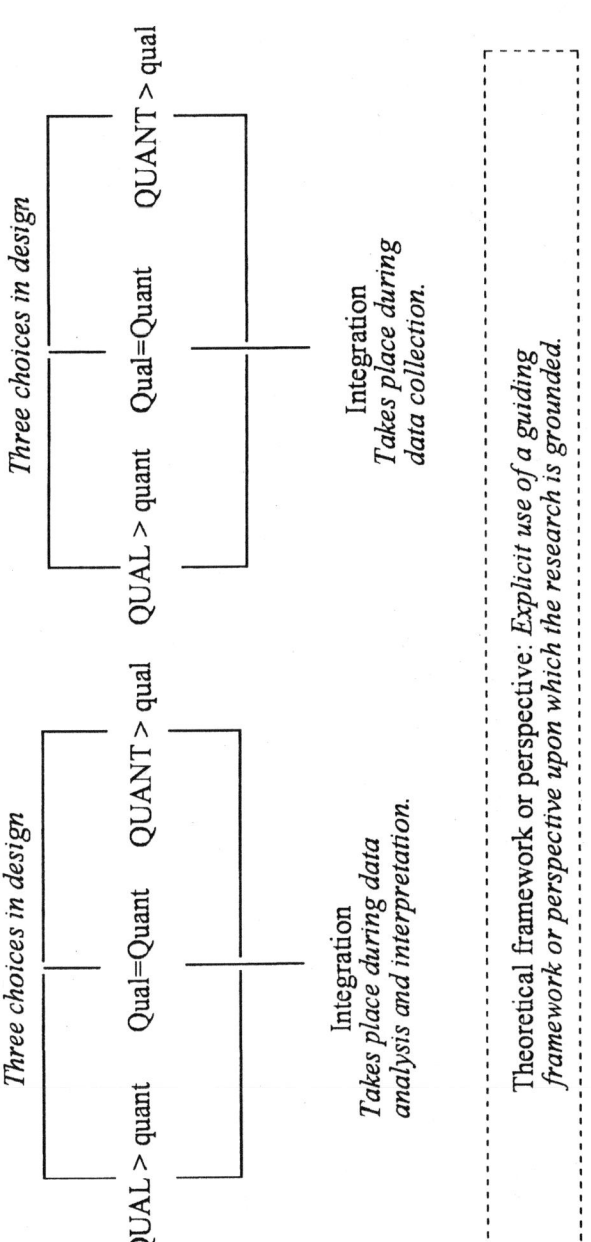

Figure 1. Key decisions and choices in mixed methods design.

study (2), the qualitative approach had priority over the quantitative approach because the qualitative data and analysis was used to formulate the final survey. In the Bunce, VandenPlas, and Havanki (4) study the emphasis on quantitative data collection and analysis indicated that the quantitative approach had priority. The theoretical framework as discussed below may also influence priority.

Integration

In designing a mixed methods study, the researcher also decides at what point to *integrate* the two approaches. It may require transforming one type of data in to another in order to integrate, compare, and analyze it. The integration process may include changing qualitative codes or categories into quantitative counts or grouping quantitative data into factors or themes. Integration may take place during data collection in a concurrent study when both open-ended and Likert-scale questions are asked on a survey. It may take place during data analysis or interpretation during a sequential study.

Theoretical Framework

Finally, as was discussed in Michael Abraham's chapter (5) on the importance of theoretical frameworks, a *theoretical perspective or framework* guides the entire research design. In a mixed methods study the theoretical framework influences the researcher's implementation, priority, and integrative decisions. Identifying a theoretical framework provides greater clarity and coherence to the proposed research and provides a lens through which the results can be interpreted.

Examples of Sequential and Concurrent Design Strategies

Researchers make decisions pertaining to each of the four factors described above to develop their mixed methods design. There are two broad categories of mixed methods implementation strategies—sequential and concurrent. Within each of these categories decisions pertaining to the dominance or equivalence of each research approach and the integration of data collection and analysis further distinguish the overall design. Although the following discussion does not exhaust all design possibilities, it highlights those most useful to those engaged in research on teaching and learning in chemistry.

The Sequential Exploratory Design

Mulford and Robinson (2) used a *sequential exploratory design* to develop the "Chemistry Concepts Inventory," (CCI), a survey of alternative conceptions for use in first-semester general chemistry. The design of the study is illustrated in Figure 2.

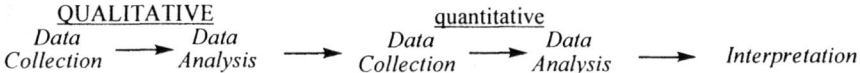

Figure 2. The sequential exploratory design strategy (Creswell, J. W. Research Design, Qualitative, Quantitative, and Mixed Methods Approaches 2nd ed. p. 213, copyright 2003 by Sage Publications, Inc. Reprinted by Permission of Sage Publications, Inc.).

The dominant research approach used in the study was a qualitative one. The initial piloted survey was crafted as a free-response instrument. The analysis of this data drove the development of the CCI. It allowed the researchers to develop a multiple-choice instrument where the answer choices reflected the students' alternate conceptions. The qualitative analysis flowed into and shaped the quantitative data collection. The CCI was given to first semester general chemistry students at the beginning and end of the course. The quantitative analysis of the results allowed Mulford and Robinson to explore the extent of student misconceptions and their robustness after a semester of instruction.

The sequential exploratory design is well aligned with survey development. As Morse (6) stated it allows researchers to explore the distribution of particular phenomena across a population. It can also be used to generalize the findings of a qualitative study to a broader population by the development of a quantitative instrument that is grounded in the data.

The Sequential Explanatory Design

The sequence of the design can be reversed, and the quantitative study may be carried out first and carry the dominant priority. This strategy is known as the *sequential explanatory design* as shown in Figure 3.

Staver and Lumpe (7) used this design to investigate students' understanding of the mole concept and its use in problem solving. In the quantitative phase they analyzed examination responses probing student understanding of the mole concept, and the relationship between the atomic or molecular mass and the

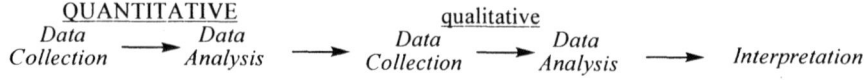

Figure 3. The sequential explanatory design strategy (Creswell, J. W. Research Design, Qualitative, Quantitative, and Mixed Methods Approaches 2nd ed. p. 213, copyright 2003 by Sage Publications, Inc. Reprinted by Permission of Sage Publications, Inc.).

molar mass. Based upon the results of the first phase, they subsequently conducted a qualitative study, which allowed them to elaborate upon the quantitative findings. Twelve chemistry students were interviewed using a think aloud interview protocol to probe their conceptual understanding of the mole and their ability to use that concept to solve problems. Analysis of the interview data revealed a frequently used misconception, specifically that the students considered the gram and the atomic mass unit to be equivalent. This numerical identity confusion was an obstacle for students as they attempted to solve mole concept problems.

The strength of this research design is that it allows researchers to elaborate, enhance, or clarify the understanding of the quantitative outcomes. For example, if the quantitative results are surprising, then the qualitative study can further examine these results from a different perspective. In the Staver and Lumpe study it allowed the researchers to clarify and explain the students reasoning, which was guided by a robust misconception (*7*).

The Concurrent Triangulation Strategy

The qualitative and quantitative data collection and analysis may be carried out concurrently as shown in Figure 4. When researchers investigate the same phenomenon or construct in both the qualitative and quantitative studies the design is known as the *concurrent triangulation strategy*. The ideal design gives the qualitative and quantitative studies equal priority (one is not dominant over the other), but in actual practice one study may dominate over the other. The use of both methods simultaneously allows researchers to counteract the biases or weaknesses in either qualitative or quantitative approaches (see chapters 7 and 8 in this text (*8, 9*) for more detail) and provides methodological triangulation (*10*). The goal is to have the findings generated by each study confirm, converge, or corroborate each other. It is a strong research design that produces well-validated outcomes and it may be the most familiar of mixed methods research designs to those interested in research on teaching and learning in chemistry.

Greenbowe and Meltzer (*11*) used a concurrent triangulation strategy to uncover the conceptual difficulties faced by college chemistry students studying

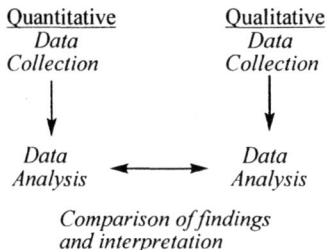

Figure 4. The concurrent triangulation strategy (Creswell, J. W. Research Design, Qualitative, Quantitative, and Mixed Methods Approaches 2nd ed. p. 214, copyright 2003 by Sage Publications, Inc. Reprinted by Permission of Sage Publications, Inc.).

calorimetry. The study used a detailed analysis of student performance on written exams (185 students on the second exam and 207 on the final, each from a pool of 541) to identify problem solving approaches and misconceptions. Multiple interviews from a single student, who the researchers identified as being "representative of a significant portion of the larger sample," were carried out during the semester and after the final (*11*). The analysis of student responses on the second exam and final were used to guide the questions posed during the interviews. Outcomes from both studies converged, allowing the researchers to recommend specific topic areas for enhanced instruction that may yield improved student understanding.

The Concurrent Nested Strategy

Similar to the concurrent triangulation approach, the *concurrent nested strategy* uses one data collection analysis phase. However, as indicated in Figure 5, one approach has priority over the other and the more dominant methodology guides the study. The less dominant method may address a different research question or collect data at different levels—teaching assistants rather than faculty or students, or administrators rather than faculty members.

Bunce, VandenPlas, and Havanki (*4*) used a concurrent nested approach to explore whether the use of a student response system (SRS) and WebCT quizzes had an effect on student achievement on both teacher written exams and an ACS standardized exam. The dominant quantitative portion of the study used a variety of measures including the GALT logical reasoning ability test, SRS scores, WebCT quiz scores, three teacher-written exam scores, and the ACS General, Organic, and Biochemistry exam form 2000 to measure student achievement. The less dominant qualitative portion of the study focused on the

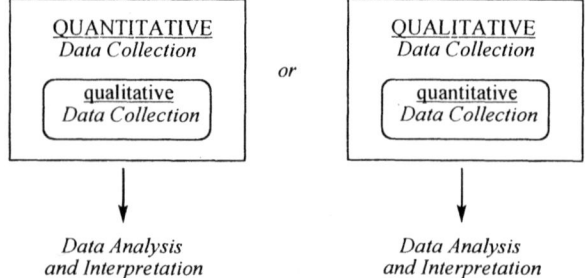

Figure 5. The concurrent nested strategy (Creswell, J. W. Research Design, Qualitative, Quantitative, and Mixed Methods Approaches 2nd ed. p. 213, copyright 2003 by Sage Publications, Inc. Reprinted by Permission of Sage Publications, Inc.).

effect of SRS use on student attitudes towards the course. A Likert-type survey and free-response questions were used to evaluate the usefulness of the SRS from the student's perspective. The analysis of the results from both approaches allowed Bunce et al. to recommend classroom practices that would lead to a more productive use of SRS and encourage meaningful learning. In addition, they also recommended that SRS questions be available for student review rather than continuing the practice of having them only available during lecture. The qualitative data allowed Bunce et al. to interpret the data more broadly and added insight and depth to their analysis.

Data Analysis and Integration in Mixed Methods Research

The previous chapters by Bretz on qualitative research (9) and Sanger on quantitative research (8) have examined methods of data analysis germane to those traditions. In a research endeavor that employs a mixed methods design, the important analysis question focuses on how the data is integrated, related, and mixed (3, 12). How is data transformed from one type into another so that the results from sequential studies may be compared? Similarly, how is data transformed during a concurrent study to facilitate interpretation?

Sequential explanatory designs can use factor analysis—the grouping of quantitative data into factors or categories—to convert quantitative data into themes. This transformation is used to guide subsequent qualitative data collection, analysis, and interpretation. It may provide areas for researchers to explore in greater depth with participants in the qualitative phase of the study.

Mulford and Robinson (2) used a sequential exploratory design to develop the CCI. The qualitative free response data was analyzed such that responses to multiple-choice questions could be derived from it. In this study the integration of the two approaches flowed from the qualitative data analysis to the quantitative data collection via the CCI. The quantitative data collection and interpretation was shaped and guided by the qualitative analysis.

In a study using a concurrent design the data collection and analysis for both methods takes place in one phase. Thus, the integration of the two methods will have a greater degree of "back and forth" rather than "flow" from one phase into another as was the case in a sequential study. However, the quantization or qualitization of the data will use many of the same procedures as a sequential study. The quantitative data can be grouped by factors (derived from the factor analysis of data) or sorted into themes and compared to the emergent findings from the qualitative study. Qualitative data can be quantized into counts, or coded numerically (see Abraham's work (*13, 14*) for example) and integrated into the quantitative analysis. Bunce, VandenPlas, and Havanki (*4*) used a different strategy, grouping the qualitative data into categories that matched the quantitative data analysis.

When designing a mixed methods research study it is important to clarify the data collection, analysis, and integration procedures at the outset of the study. A well-designed data analysis plan can remove the danger of being left awash in data. Multiple forms of data and analysis require very specific research designs including careful consideration of how the data is to be transformed and integrated.

Examples of Mixed methods Strategies in the Chemistry Education Literature

Articles from the *Journal of Chemical Education*, and the *Journal of Research in Science Teaching* have been selected to illustrate mixed methods designs in chemical education research. Representative publications listed below serve as resources for those interested in designing and conducting mixed methods research. The methodological detail in these references allows the reader to identify the data collection instruments (ACS examinations, surveys, interview protocols, etc.) and analysis techniques in most cases.

- Staver, J; Lumpe, S. Two Investigations of Students' Understanding of the Mole Concept and Its Use in Problem Solving. *J. Res. Sci. Teaching*, **1995**, *32*, 177-193 (*7*). The data collection and analysis methods in both the quantitative and qualitative studies is explained in detail. This study is an excellent example of a sequential explanatory design.
- Gutwill-Wise, J. P. The Impact of Active and Context Based Learning in Introductory Chemistry Courses: An Early Evaluation of the Modular

Approach. *J. Chem. Ed* **2001**, *78*, 684-690 (*15*). The article reports two comparative studies that assess the impact of the Modular approach on students understanding, reasoning skills, and attitudes toward chemistry. The supplemental materials provide the reader with access to the conceptual tests, attitudinal surveys, and interview coding schemes. There is an abundance of methodological detail in this article that provides readers with the tools used to collect and analyze data.

- Donovan, W. J.; Nakhleh, M. B., Students' Use of Web Based Tutorial Materials and Their Understanding of Chemistry Concepts. *J. Chem. Ed* **2001**, *78*, 975-980 (*16*). The authors collected and analyzed data using a survey that included scaled-response and free-response questions. In addition, a small group of students was interviewed.

- Mulford, D. R.; Robinson, W. R. An Inventory for Alternate Conceptions among First-Semester General Chemistry Students, *J. Chem. Ed* **2002**, *79*, 739-744 (*2*). This article nicely documents the development of this inventory using a sequential exploratory research design.

- Herrington, D. G; Nakhleh, M. B.; What Defines Effective Chemistry Laboratory Instruction? Teaching Assistant and Student Perspectives. *J. Chem. Ed* **2003**, *80*, 1197-1205 (*17*). The article includes the entire survey that included one free response question. The findings from the qualitative analysis of the free response data are very carefully explained. An inter-rater reliability study is included.

- Teichert, M. A.; Stacy, A. M.; Promoting Understanding of Chemical Bonding and Spontaneity Through Student Explanation and Integration of Ideas *J Res Sci Teaching*, **2003**, *39*(6), 464-496 (*18*). Quantitative achievement data and interview data are included. The worksheets used during the interventions are included, but the modified coding scheme and interview protocol are not.

- Bunce, D. M.; VandenPlas, J. R.; Havanki, K. L.; Comparing the Effectiveness on Student Achievement of a Student Response System versus Online WebCT Quizzes. *J. Chem. Ed* **2006**, *83*, 488-493 (*4*). The methodology used on both the quantitative and qualitative studies is thoroughly explained. This is an example of a concurrent nested research design where the quantitative study had priority over the qualitative study.

In order to contribute to the development of mixed method strategies in the chemical education community it is imperative that authors provide sufficient methodological detail about both quantitative and qualitative methods. These details must be retained through peer review and publication so that others may learn from these studies.

Conclusion

Pragmatically, mixed methods designs hold the promise of explaining or exploring the phenomenon under investigation with greater depth and breadth than choosing a design with one research strategy. In a mixed methods study the researcher uses a more complex analysis integrating the data to achieve greater interpretability than could be achieved using a single research approach. The qualities of outcomes from quantitative studies—generalizability and statistical reliability—and qualitative studies—rich, thick description—are achievable with mixed methods designs. Ideally these designs will lead to new relationships, generate new insights, and develop findings from a wider range of perspectives that will be constructive and useful to chemistry faculty and chemistry education researchers.

References

1. Sanger, M.J.; Badger, S.M. *J. Chem. Educ.* **2001**, *78*, 1412-1416.
2. Mulford, D. R.; Robinson, W. R. *J. Chem. Educ.* **2002**, *79*, 739-744.
3. Creswell, J. W. *Research Design, Qualitative, Quantitative, and Mixed Methods Approaches* 2nd ed. Sage: Thousand Oaks, 2003.
4. Bunce, D. M.; VandenPlas, J. R.; Havanki, K. L *J. Chem. Educ.* **2006**, *83*, 488-493.
5. Abraham, M.R. In *Nuts and Bolts of Chemical Education Research*; D. Bunce and R. Cole (Eds.); American Chemical Society Symposium Series, Washington, D.C.: 2007.
6. Morse, J. M. *Nurs. Res.,* **1991**, *40*, 120-123.
7. Staver, J; Lumpe, S. *J. Res. Sci. Teaching,* **1995**, *32*, 177-193.
8. Sanger, M.J. In *Nuts and Bolts of Chemical Education Research*; D. Bunce and R. Cole (Eds.); American Chemical Society Symposium Series, Washington, D.C.: 2007.
9. Bretz, S.L. In *Nuts and Bolts of Chemical Education Research*; D. Bunce and R. Cole (Eds.); American Chemical Society Symposium Series, Washington, D.C.: 2007.
10. Patton, M. Q, *Qualitative Evaluation and Research Methods*, 3rd ed.: Sage: Newbury Park, 2001.
11. Greenbowe, T. J.; Meltzer, D. E. *Int. J. Sci. Educ.* **2003**, *25*, 779-800.
12. Frechtling, J., Sharp, Laure, (Eds.). *User-Friendly Handbook for Mixed Method Evaluations*. National Science Foundation: Arlington, VA, 1997.
13. Haidar, A. H. Abraham, M. R. *J Res Sci Teaching* **1991**, *28*, 919-38.

14. Williamson, V. M.; Abraham, M. R. *J Res Sci Teaching* **1995**, *32*, 521-534.
15. Gutwill-Wise, J. P. *J. Chem. Educ.* **2001**, *78*, 684-690.
16. Donovan, W. J.; Nakhleh, M. B. *J. Chem. Educ.* **2001**, *78*, 975-980.
17. Herrington, D. G; Nakhleh, M. B.; *J. Chem. Educ.* **2003**, *80*, 1197-1205.
18. Teichert, M. A.; Stacy, A. M. *J Res Sci Teaching*, **2003**, *39*, 464-496.

Chapter 10

Designing Tests and Surveys for Chemistry Education Research

Kathryn Scantlebury[1] and William J. Boone[2]

[1]Department of Chemistry and Biochemistry, University of Delaware, Newark, DE 19716
[2]Ohio's Evaluation and Assessment Center, Miami University, Oxford, OH 45056

This chapter will provide readers with an overview of how to design and evaluate surveys and tests using 1) pencil and paper techniques and 2) the Rasch psychometric model. The goal is to help chemistry education researchers develop robust tests and surveys that optimize data collection. For paper and pencil test development, we discuss issues such as the importance of item wording, the use of figures, how to best select a rating scale and the impact of text layout. Then, we present an overview of how researchers can use Rasch analysis to 1) guide the initial development of instruments, 2) evaluate the quality of a data set, 3) communicate research findings and 4) carry out longitudinal studies. The careful development of measurement instruments, in addition to the use of Rasch techniques can help optimize what is learned in many chemistry education studies.

Introduction

There are many issues of common concern when one develops tests and surveys. A researcher must decide to pool items for an overall "measure" or to individually examine items and the type of rating scale in an attitudinal survey. In this chapter, we will present some critical issues, which researchers should consider when developing surveys and/or tests for data collection and describe how researchers can use a psychometric model (the Rasch model) to 1) improve the development of tests and surveys, 2) evaluate survey and test data, and 3) communicate research findings. The first part of the chapter can be viewed as being "paper and pencil" in nature, while the second part of this chapter should be viewed as a critical step in the development and analysis of test and survey data. However, researchers also need to respect the steps in test development for the Rasch model to be useful. That is, the Rasch model cannot compensate for poorly designed tests and surveys. We have chosen to organize both the "paper and pencil" and Rasch portion to provide a user-friendly overview of critical issues when designing tests and surveys. Some design issues impact only tests while other issues impact surveys. We have attempted to present a range of issues in this chapter which will aid researchers designing both tests and surveys, however to minimize chapter length we do not present an exhaustive discussion of all possible issues.

Writing Surveys and Tests

Pooling Items

When researchers design a survey or test, it is important to consider whether or not there is a goal of "pooling" a set of items to produce an overall measure. To understand this issue, consider what is gained when a test of numerous items is presented to a student. If a student completes a 50-item test, then there are 50 items to determine her/his performance. Teachers commonly recognize this as the calculation of the student's total test score that is a more precise assessment of a student's performance than the student's performance on single test items. It is important to mention that "pooling" items together in an effort to increase measurement precision should not be carried out without considering important reliability and validity issues. For example, a chemistry test written to assess knowledge in biochemical principles would be inappropriate if all the items were focused on physical chemistry topics. Similarly, giving a chemistry test designed for graduate students to first year undergraduates would be unfair. Thus, test developers need to carefully consider which items on a 50-item test would "pool" together to measure different aspects of a student's chemical knowledge.

If the goal is to "pool" items, there are specific techniques to design a test, for example, the respondents' breadth and range of knowledge. To achieve this goal, a researcher would develop test items with a mix of difficulty (e.g. easy, medium and hard) – this helps differentiate the performance of the respondents. We have found when authoring items for a test, it is useful to predict the difficulty of test items. Figure 1 presents a fictitious five-item test and the location of predicted item difficulty - predicted by the item author.

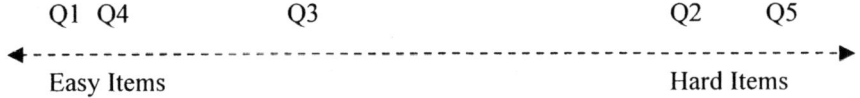

Figure 1. Prediction of item difficulty by a test item author.

Authoring items and predicting item difficulty forces test item authors to think in more detail about test development. Secondly, by predicting item difficulty, item authors can learn if they might have authored too many, or too few, items at one difficulty level. Third, when data is collected, researchers can compare predicted to actual item difficulty placement. This serves as a technique to improve test item authors' understanding of the issues they are investigating. For instance in Figure 1, if item Q1 is more difficult than predicted, it may suggest a problem in the item's structure. Test item authors can use the mismatch between predicted and actual item difficulty as an indicator to review and revise items.

Test item authors can also use these steps for authoring attitudinal survey items. For instance, if a respondent can indicate that they "agree" or "disagree" with a statement. One should include -

- Items that are easy for respondents to select "agree" and
- Items that are difficult to agree with.

When researchers develop a wide range of attitudinal items, they maximize the differentiation between the attitudes of respondents, similar to tests containing a wide range of items. Figure 2 presents a schematic that displays a possible distribution of predicted ease with which respondents agree when the researcher uses five survey items.

A researcher may have a large and varied number of project goals and as such, a survey or test need not always contain pooled items. For example, in some circumstances researchers might only administer a brief survey or test to respondents and in such a case, it would be too onerous to present a number of test or survey items for each goal. The types of issues presented in Figures 1 and 2 are discussed in *Best Test Design (1)*.

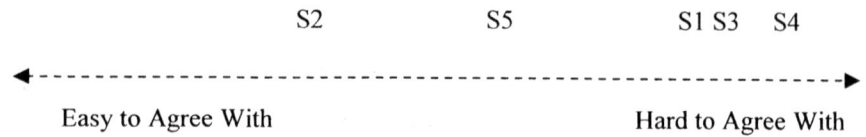

Figure 2. Schematic representing range of attitudinal items.

Quality Figures Yield Quality Items

Other issues also affect item quality. For example, if an item includes a figure or a picture, it must be clear what the figure represents. A caption helps to provide textual clarity to figures. When figures are placed in tests, their visual quality and clarity are important. The consistent use of figures is also important. For example, if a test item includes a picture of a burette, then it is important to utilize the same picture in subsequent test items with burettes. We have found it useful to have potential test/survey takers review figures before using the final test for data collection.

Item Style and Word Choices

Are You Planning to Underline or Bold Selected Words?

Authors' word choices, for both the stem (the question) and answers, are crucial for the development of high quality test items. The same is true for text in attitudinal surveys. Occasionally in test and survey items, selected words are underlined, bolded, or capitalized. These are actually style choices, not hard-and-fast rules. We suggest that researchers be consistent with respect to the "rules" they use for item authoring. Thus,

- If the researcher decides that "key" words will be underlined, then it is necessary to underline all "key" words.
- If the author decides that the word *all* will be underlined in a survey or test item, then *all* should be underlined in each item in which it appears.

Are You Encouraging Data Quality Problems by Using "And" or "Or"?

Survey and test items may include words such as *and* or *or*. For improved tests and surveys we suggest that researchers exclude these words from test/survey items. For example, consider the statement,

"I learned from lectures and labs in this class."

The word *and* plays a critical role in respondents' answers. Respondents agree with this item only if they learned from both lectures and labs. Authors should split the item in two:

- "I learned from lectures in this class" and
- "I learned from labs in this class."

The word *or* is also problematic. Consider a survey item that states,

"Next semester I would like to enroll in an independent study chemistry class or an organic chemistry class."

When students agree with this item, we simply learn that they agree with plans to enroll in at least one of the two classes mentioned. We realize that researchers juggle pragmatic issues, such as a limited number of items that can be presented to a survey respondent or a test taker. However, we simply caution the survey or test designer to think of the implications of utilizing words such as "and" or "or".

What are you testing? What is the goal of your survey? What issues can diminish your success in optimally testing or optimally surveying?

In the previous section involving the "pooling" of test items, we briefly discussed issues associated with which individual items on a test might be utilized for a total measure. Perhaps, one of the most important aspects of survey and test design is the development of items that involve one construct. A construct should be viewed as a single key variable of interest. Examples of every day constructs are length and weight. When developing a measurement device it is important to author items so that one key issue is addressed by all survey items or test items that will be pooled together. This type of issue has been discussed in great detail by Benjamin Wright, of the University of Chicago, in his numerous publications *(1-2)*. Wilson *(3)* can also be consulted for guidance on these and other issues.

There are many additional factors that affect the strength of a test or survey beyond the consideration of a construct. Generally, the researcher minimizes issues that are tangential to the goal of a test or survey. For example, consider the role that *reading* plays as a respondent answers test or survey items. Most test and survey items should not depend solely on vocabulary knowledge. For example, consider an item that a researcher is planning to present to 4th grade elementary students following a museum field trip. The possible survey item might be: *Did the museum program help you better comprehend how chemistry is used to make paint?* This survey item is not only long, but the word "comprehend" might be too difficult for many of the students. It might be much better to present the word "understand". When developing tests and surveys, it is best to minimize issues of no interest. For instance, if the goal is to determine how much chemistry a student knows, then researchers should minimize the issue of reading, ensuring the test is solely a chemistry test.

What is the Preferred Length of an Item?

When authoring items, short and sweet is better than long and drawn out. Shorter items allow respondents to more easily (and more quickly) complete a survey. For each line, respondents must read from left to right, the final line is a set of responses, read from left to right. For example, consider the following survey item:

"In reviewing the curriculum materials developed for this class, I found that the text was well organized."

In this case the reader must read from left to right, and at the end of the line they must read from left to right again. This subtle difference allows a reader to process the text more easily. Shorter items decrease the amount of time respondents need to mentally process the item, while reducing the potential for respondent frustration due to lengthy surveys. Researchers also benefit from the likelihood that respondents may spend more time thinking carefully about their answers.

Construct Quality Test Distractors

Multiple-choice tests present respondents with a sequence of questions with a number of possible answers. These choices are often termed "distractors". Tests should include the same number of distractors for each multiple-choice question that include viable (believable) incorrect answers. Thus, respondents may be distracted from the correct answer. Consistently writing viable distractors is a difficult task, but "give away" items increase the respondents' chances of a correct guess. The goal is to test the respondents' content knowledge, not the ability to guess correct answers on multiple-choice tests. Occasionally, researchers write test answers that include the phrases "all of the above" or "none of the above". The usefulness of such phrases is greatly dependant upon the quality of other feasible responses presented to test takers.

Changing Item Wording to Match a Rating Category

Another issue is the respondents' comprehension of the rating scale. Consider a project in which chemistry teachers rate the importance of a particular teaching component for student learning. The item below might be presented to teachers with the following rating scale category:

"Please indicate how important the following issues are with respect to enhancing student learning in chemistry laboratories."

1) Group work

Very Important	Important	Unimportant	Very Unimportant

2) Lab periods of at least 2 hours in length

Very Important	Important	Unimportant	Very Unimportant

The rating scale is understandable, but using "very unimportant" could be problematic as it is an uncommon phrase. One technique is to alter the items phrasing to use an "agree" rating scale. Below we present the revised directions to teachers, the altered questions posed to teachers, and the new rating scale:

"Please indicate your level of agreement with respect to the following statements concerning how to enhance student learning in chemistry laboratories."

1) Group work is an important strategy to improve chemistry laboratory learning.

Strongly Agree	Agree	Disagree	Strongly Disagree

2) Lab periods of at least 2 hours in length is a strategy to improve chemistry learning.

Strongly Agree	Agree	Disagree	Strongly Disagree

There are no set "rules" that survey developers must follow regarding the phrasing of survey items and the selection of rating categories. However, careful rephrasing of items may allow researchers to use a more understandable and less awkward rating scale. This improves the quality of data collected.

Utilizing an Even or Odd Number of Rating Categories

There are some rating scales for which an odd or even number of rating categories is unimportant. A researcher may select one rating scale to identify the frequency of an activity or behavior, such as very often, often, sometimes, seldom, never. However, sometimes respondents may indicate that their responses are exactly half way in-between the two extreme rating categories. Compare the following two rating categories:

Agree	Disagree

Versus

Agree	Neither Agree nor Disagree	Disagree

Some researchers advocate the provision of a middle category for respondents. However, presentation of a middle category can allow respondents to ignore some of a survey's items. The respondents' selection of the "Neither Agree nor Disagree" category may indicate they are quickly answering a survey item. It is best to avoid phrases such as "not sure," "neutral" or "neither agree nor disagree".

Reversed Items (Or "to flip or not to flip")

In many surveys or tests, it is important to receive thoughtful responses from respondents. For example, undergraduate students might be presented with the following survey items.

1) I like to complete chemistry labs.

Strongly Agree	Agree	Disagree	Strongly Disagree

2) I like to work with a lab partner.

| Strongly Agree | Agree | Disagree | Strongly Disagree |

3) I do not like to use lab equipment.

| Strongly Agree | Agree | Disagree | Strongly Disagree |

4) The labs will enable me to do well in subsequent chemistry classes.

| Strongly Agree | Agree | Disagree | Strongly Disagree |

The idea, in terms of survey construction, is that item 3 would be one that would keep respondents on their toes in terms of carefully reading an item. The assumption is that for someone who answers "strongly agree" to items 1, 2, and 4 - that the same person would most probably answer, "strongly disagree" to item 3. The thought among some test and survey construction experts is that item 3 would remind survey takers to carefully read each item. But we suggest there are some possible problems with the use of the word NOT. The usefulness of item 3 depends upon survey respondent's careful reading of an item.

Consistent wording, consistent phrasing

A key issue that is often overlooked in the development of both tests and surveys involves the use of consistent phrases from item to item. Consider the following sample items:

1) This lab experience helped me learn chemistry topics.

| Strongly Agree | Agree | Disagree | Strongly Disagree |

2) The lectures for this class aided my comprehension of chemistry class material.

| Strongly Agree | Agree | Disagree | Strongly Disagree |

3) Class topics were much easier to understand as the result of discussion sections.

Strongly Agree	Agree	Disagree	Strongly Disagree

Each of the items attempts to collect student attitudinal data on three components of a chemistry course (labs, lectures and discussion sections). But the structure and the wording of items do not optimize the certainty with which researchers can compare student responses from item to item. A few of the problems are illustrated by the following example set.

Each item begins with a different type of lead-in (This lab experience, The lectures, Class topics). An improvement would be either of the following options:

Option 1- The lab experiences…, The lectures…, The class topics…
Option 2- Lab experiences…, Lectures…, Class topics….

There are always key words that tilt the meaning of a survey item. In this example, three different words are used to indicate learning:

Q1- learn… Q2- comprehension… Q3 - understand…

It is important to carefully monitor the key words used within a survey (or test) item. In this case, better instrument construction will result from the researcher's use of only one of the three words, *learn, comprehend*, or *understand*, in appropriate items. A similar problem occurs with the use of the words *helped* (Q1) and *aided* (Q2). In order for a researcher to remain consistent, both items (Q1 and Q2) should utilize only one of the two words, either *helped* or *aided*.

Finally, words that can intensify other words in a statement must be carefully maintained (For example, Q3 includes the phrase "much easier"). Often, researchers are inconsistent with the use of key words such as "much," "very" and "some". We return to our previous set of questions to demonstrate suggested revisions; examples of revised items are presented below and illustrate how edits can improve items and data collection.

1) Lab experience in this course helped me learn class topics.

Strongly Agree	Agree	Disagree	Strongly Disagree

2) Lectures in this course helped me learn class topics.

Strongly Agree	Agree	Disagree	Strongly Disagree

3) Discussion sections in this course helped me learn class topics.

Strongly Agree	Agree	Disagree	Strongly Disagree

Physical Layout of the Instrument

In previous sections of this chapter, we have discussed aspects of test and survey layout. The time investment to produce a visually appealing layout improves the data quality and minimizes problems caused by respondents' confusion, misinterpretation, or misunderstanding of items. Some suggestions for improving instrument layout include:

- Limit the amount of reading required for any single item. Longer items often involve more than one issue.
- Carefully work font, spacing, and indentation to improve a survey's legibility.
- Do not let items run from one page to another.
- Number the pages.
- Present fewer items in larger more easily readable font rather than more items in a small, hard-to-read font.
- Include a cover page for respondents' anonymity.

Writing Short Items (Revisited)

Survey and test designers should consider survey length and avoid asking respondents' questions that may not be used in data analysis. The strategy of minimizing the time a respondent needs to answer the survey will improve data quality because of reducing respondent fatigue. For example it might be tempting to ask respondents to provide added detail with respect to an issue such as "past courses completed". However, respondents' additional information will limit the amount of time that they spend on the survey. If the data is unanalyzed then remove the items from the survey because it takes time away from the respondents, and the data takes time for researchers to enter into databases.

Piloting of Instruments

It is important to proof read and pilot instruments. The first step in the piloting process is to administer the instrument to 20-25 respondents who are similar to those in the sample population. These test respondents should be asked to complete and write comments about the survey (or test). The comments may involve critiquing directions and rating scales, and suggesting revisions to item wording. Second, test developers can use the respondents' answers to identify data patterns. For example, respondents may not use all of the attitudinal rating categories. Such a response pattern may suggest that the selected rating category will not be sufficiently precise to separate the attitudes of respondents. For example, consider the three items presented earlier in this discussion. If the pilot group only selected either Strongly Agree or Agree, then the scale acts a two-step rating scale that limits the instrument's precision. In this situation, researchers have options:

Option 1 - Wait and see if such a pattern persists with the real data set.

Option 2 - Alter the rating scale. In our example, one might change the rating scale to be the following:

1) Lab experience in this course helped me learn class topics.

Strongly Agree	Agree	Barely Agree	Barely Disagree	Disagree	Strongly Disagree

Such a change might help one elicit a wider range of attitudinal selections made by respondents.

Option 3 - Alter the wording of survey items so that more of the rating scale might be used.

1) Lab experience in this course helped me easily learn class topics.

Strongly Agree	Agree	Disagree	Strongly Disagree

Option 4 - Alter the wording of survey items and use a different rating scale.

1) Lab experience for this course helped me easily learn class topics.

Strongly Agree	Agree	Barely Agree	Barely Disagree	Disagree	Strongly Disagree

Adapting Existing Instruments

What to Assume, What Not to Assume

Using previously developed instruments sometimes provides researchers with easier data collection and the ability to compare results across settings. There are caveats when using existing instruments. One assumption is that a peer-reviewed, published instrument is valid and reliable, but often researchers do not provide the stages of instrument development, piloting, field-testing, and psychometric analysis. While we encourage researchers to use existing instruments, good research practice includes completing psychometric analyses (e.g, factor analysis, reliability) to monitor and improve instruments. Rasch analysis is one approach researchers can use to achieve this goal.

Check Coding

Raw survey data is prepared for further analysis, hand-entered or scanned electronically into electronic files. If data is hand-entered, researchers should develop a data-coding scheme (e.g., demographics, survey items, open ended responses) during the pilot study to assist with an efficient data entry process. Researchers should also hand check 5-10% of the sample as a quality control of the data entry accuracy. Electronic scanners also have physical limitations that can constrain survey formats. Researchers should decide before producing the survey if they will use electronic scanners for data entry. The survey format should adhere to the scanners' restrictions.

Check Keys

The previous discussion on developing surveys also is applicable to developing content tests. In addition, tests should include answer keys developed by "experts" in the field. All experts reviewing a test should agree regarding the correct answer for each test item. Several individual experts should complete a test and explain their answer selections. If experts disagree on an item's answer, the item may be removed from the test. Once a key is developed, it must be checked. One method of checking a key is to include an imaginary person in the data set who correctly answered all test items. When the analysis is run, ensure the imaginary person receives a score of 100%.

Codes and colors may be used to track data

It is helpful to utilize a few methods to track multiple data sets within a study. These methods include:
- Requesting some sort of code that can be used to track respondents (e.g., initials and birth date),
- Inserting a code into the base of each survey page (e.g. Fall 2007)
- Using color-coded instruments.

For example if a survey is given at the end of Fall 2007, that survey could be printed on green paper. But if that same survey is administered in the Spring of 2008, it would be printed on gold paper.

The first part of this chapter has presented what we call "paper and pencil" tips regarding a variety of issues that can be considered when developing surveys and tests. One step in survey or test design is to compute a reliability statistic, such as Cronbach alpha. Typically, researchers collect data with their instrument and commence their statistical analysis. However, researchers should include a second analysis step that utilizes psychometric theory to guide the development of surveys and tests, to provide techniques that allow the functioning of surveys and tests to be monitored and improved, and to prepare data for statistical calculations. In the second part of the chapter, we provide an overview of one psychometric technique (the Rasch model) that allows researchers to easily carryout this important second analysis step.

Rasch Modeling

In this section, we will provide a brief overview of Rasch analysis - How it can be used to develop tests and surveys. - How the technique can be used to improve tests and surveys. - How the technique can be used to prepare data for analysis. This summary is meant to provide a simple overview. For a more thorough understanding, we encourage readers to consult texts such as *Best Test Design* (*1*), *Rating Scale Analysis* (*2*) and *Applying the Rasch Model Fundamental Measurement in the Human Sciences* (*4*), for additional details with respect to the psychometric technique.

Danish mathematician George Rasch developed the Rasch model (*5*). Researchers can use the Rasch model to develop tests and surveys, monitor the quality of survey or test data, improve test or survey items, and calculate an "equal interval" total score for both test takers and survey respondents. When researchers evaluate data using parametric statistical tests (e.g., t-test, ANOVA), they assume that *score data* is "*equal interval*". We can use the Rasch analysis software to convert "non-equal interval" data into "equal interval" data. In recent years, evaluators have used the Rasch model for large-scale, assessment projects such as the evaluation of reform in the Chicago Public Schools (*6*)

Item Maps

Figure 1 showed how item authors could predict the difficulty of test items, in order to develop a test (or survey) with a range of items. Rasch analysis software produces "item maps" which allow researchers to evaluate the distribution of items on a test from easy to hard and to identify possible "gaps" in item difficulty. An optimum test will have items with various difficulties rather than many items of the same difficulty. Researchers can also use item maps to communicate how different groups (e.g., girls and boys) perform on particular test measures. Figure 3 shows an example of an item map.

Researchers can also construct item maps and interpret survey data in a similar way. Item maps allow the researcher to evaluate how well their measurement instrument (their test or survey) is performing, as well as a visual image of the ordering of summarized data. The maps can be read (and explained) as if one were viewing a thermometer. Since thermometers are a common measurement instrument, even those individuals who might not be familiar with Rasch item maps can easily understand such displays of data. Figure 3 displays an item map for a fictitious six-item test. A value of "0" on this map indicates a very easy item, while a value of "1,000" indicates a very difficult item. What can researchers learn from this very simple map? First, items 2 and 6 and items 3 and 5 have similar "item difficulty". If a test goal is to differentiate test takers, then it would be better to remove either item 2 or 6, and 3 or 5. Researchers could substitute two new items that "fill in a missing gap". For example, a test-developer may author two new items that fall above a value of 750. For this item map, we have not provided the text of these fictitious items. In a real data set, providing the question text next to the item on the thermometer assists in test revision. For example, we can use the item map to begin the interpretation of the data set, using questions such as: What items are difficult/easy? And why is that the case?

Rasch software (7) not only easily constructs item maps but the software can also present "person measures" and items on the same map. To limit, the amount of information which has to be digested we have chosen to not include "person measures" in Figure 3. Readers can review Rasch articles in the science education literature to see such person item maps and also to be provided with interpretive details of such maps.

Item Fit and Person Fit

Rasch analysis also provides many useful statistics for researchers to monitor the quality of respondent answers and items. As with item maps, we will provide an overview of this Rasch tool. Particularly useful Rasch statistics are *item fit* and *person fit* statistics. The person fit statistic helps researchers quickly identify individuals with idiosyncratic responses. The item fit statistic

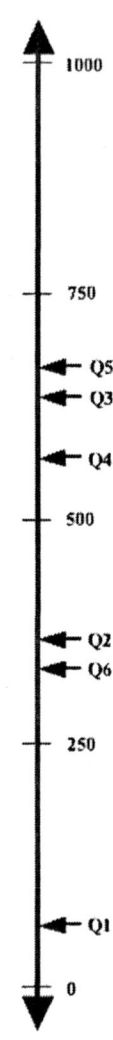

Figure 3. Example of an item map.

also helps one identify items that may cause individuals to react in an idiosyncratic manner. Statistics that indicate idiosyncratic responses or individuals are called *misfits*. There are several reasons for person "misfit" (when a person reacts to an item in an unpredictable manner). For example, a person's attention may change during the test, a miscoding on the part of a respondent (if a bubble sheet is used) or errors occurred in data entry. Another reason for person misfit may be specialized knowledge on the part of a respondent taking a test. That is, a respondent may have insightful knowledge on a particular item that may explain why s/he overall performs poorly on a test, yet unexpectedly answers one particular item correctly.

Researchers can also use item "misfit" (when an item seems to cause idiosyncratic responses) to improve a test and to evaluate a data set. An item with "misfit" may be an easy item unexpectedly missed by a number of very high performing students (those students who correctly answer almost all of a test's items). A different scenario can also cause item misfit: A test may include a difficult item for most test takers, but a number of students who did very poorly on the test may correctly answer the item. In this case, item "misfit" may also indicate a problem in item wording.

Rasch analysis of a data set provides a range of diagnostic statistics. As was the case for item maps, it is not difficult to calculate Rasch fit statistics for both respondents' survey items and test items. Review of such statistics allows a targeted monitoring of both "how" items function, as well as "how" individuals' answer a survey or test. Fit statistics allow researchers to quickly monitor the quality of a data set, and flag problematic issues. For example, 1) Do particular types of items cause misfit? 2) Do particular subgroups of students exhibit misfit? 3) Does the location of a test item (at the start of a test or end of a test) predict item "misfit"?

Anchoring

In Rasch analysis, "*item anchoring*" allows researchers to collect data using similar but not identical forms of a test or survey over the course of a project. With *item anchoring*, we can use the Rasch software to link similar forms of a test or survey. This ability to link forms at many time points provides great flexibility for researchers. In some cases, a researcher may want to administer a test to one group of students, and administer a similar version to another group of students. Since the tests are not identical, then it would be incorrect to immediately assume that a raw score for one test (say 15/20 on Form A) had the same meaning as an identical score on the Form B test. The two tests, even if many items are identical, may not be identical in overall average difficulty. Rasch analysis can ensure that the tests have identical average difficulty.

Researchers can use *anchoring* over the course of a project to modify a test, yet, all respondents are ranked on the same measurement scale. For example, researchers developed Form 1 and collected one year of data collection from over 1,000 students. When revising the test using item maps, the test developers remove some items and insert new items, thus producing a new test (Form 2). By retaining some common items from the Form 1 to Form 2, it is possible to express the performance of test takers completing the new test on the same scale as that used for the old test. The ability to link old and new forms of a test over time provides great flexibility to the researcher.

Evaluating Rating Scales

The methods provided above (*item maps, fit statistics, anchoring*) detail how researchers can utilize the Rasch model to evaluate tests and surveys. There are many positive aspects of using the model that benefits those involved in the collection and analysis of survey data. In the first part of the chapter, we briefly discussed advantages of piloting the collection of survey data with a small sample of respondents. With the collection of pilot data, one can revise a rating scale. Rasch analysis of data provides a wide range of user-friendly techniques to improve the collection of survey data using a rating scale. Basic Rasch diagnostic plots allow researchers to evaluate the functioning of a rating scale. Combining of rating scale categories may improve the scale. For example, a rating scale may initially utilize rating categories of *strongly agree, agree, barely agree, barely disagree, disagree,* and *strongly disagree*. However, data analysis may reveal that combining the *agree* and *barely agree* categories and the *disagree* and *barely disagree* categories provides better results.

Raw Score-Equal Interval Conversion Table

By conducting a Rasch analysis of basic test (or survey) data, researchers can quickly convert possibly non-equal interval raw data to interval data. To best understand this issue, consider the following: Sarah completes a test and earns 95/100, while Pam earns 90/100. The difference between Sarah and Pam may not necessarily be the same as the difference between Tom who earns 50/100 and Henry who scores 55/100. Although the differences (5 points in each case) between the raw scores are the same, the raw score at different parts of the scale may not have the same substantive meaning. Rasch analysis software can be easily used to convert the possibly non-interval raw score data of tests and surveys to an interval scale, and it is that data which is used for statistical analyses.

Software

Several software programs support Rasch analysis. Winsteps (7) is user-friendly and the author provides technical support via email. Numerous Winsteps workshops are regularly offered throughout the world. Below we provide a very simple "control file" that can be used to evaluate a ten-item, multiple-choice data set. We have provided comments throughout the sample control file. Our intention is to show readers a simple Rasch analysis with a small amount of understandable code. The Winsteps software site also provides free Rasch software called Ministeps (7). Researchers can use Ministeps for a Rasch analysis of data.

```
;This line is for the title on each page
TITLE=' A 6 item test'
;
;The data is one line per person, first a 10 column id for each person and
;then their letter answers to the 6 item multiple-choice test
FORMAT=(10A1,6(1A1))
;
; The first column of data is the first letter of the 10 character ID
NAME1=1
; This line just tells Winsteps how long the ID is
NAMELEN=10
; This line just tells the program that the 11th bit of information read in is the
;1st test item
;ITEM1=11
; This tells the program there are a total of 6 test items
NI=6
; This is the answer key
KEY1=caadbb
; This tells the program all allowed answers for all items
CODES="abcde"
;The line below end part of the code, it is followed by a short description of
; 6 items and then that is followed by 5 lines of sample data - first the
; person id followed by the item answers.
&END
Q1-Physical properties
Q2-Animal characteristics
Q3-Controlling variables
Q4-Inertia
Q5-Controlling variables/experimental design
Q6-Experimental design
END NAMES
2543305239ccbacd
```

1912203179cacbac
1913308289cdacad
2544508319aaacdd
2545603249caadad
2547770319aabccb
2543103283caacad
2538909169cbacbe
2548707279caaabd
2545511219caadbd

Summary

In this chapter, we presented key issues for researchers to consider when designing, validating and using instruments for quantitative data collection. First we discussed issues related to instrument design such as pooling items, predicting item difficulty, and the importance of producing quality figures and pictures in items. Researchers should also consider the format, wording, item length, presentation and use of scales and number of response categories. Experts in instrument development suggest that questions should have a consistent format. The same figure or picture for an item, such as a burette, be used consistently throughout the instrument to reduce confusion among respondents. Moreover, instruments should be developed and then pilot-tested. A pilot study provides the opportunity for researchers to identify problems with an instrument before data collection.

The careful design of instruments can assist researchers in collecting data that is valid and reliable. For example, removing extraneous information and reducing text can minimize the impact of respondent fatigue in answering questions. Shorter, simpler statements can clarify the survey's purpose. Researchers should also take measures to reduce errors in hand entering data and/or using electronic scanners to produced datasets or databases. With both approaches, at least five percent of the survey answers should be hand-checked for accuracy in data entry. Another strategy for cross-checking the accuracy of a data set is to include an imaginary person who correctly answers all items.

The second section of the chapter provided an overview of the advantages of using the Rasch model for monitoring, revising and refining instruments and performing data analysis. Rasch is a probabilistic model that produces equal interval data and predicts person and item statistics. Researchers can use those statistics to identify idiosyncratic test or survey items. For example, item maps provide a visual image of item difficulty. Researchers can use this information to remove and/or edit items, or add new items to an instrument. *Item anchoring* allows researchers to modify instruments, while comparing cohorts of students. This characteristic of the Rasch model is particularly useful when evaluating on-going projects.

While, paper and pencil techniques can be a very important step for the design of surveys and tests, researchers should consider both types of technology. Free software for using the Rasch model is available and software developers provide electronic technical support. However, developing good instruments is time-consuming and researchers should consider using published instruments that have detailed psychometric information, such as their development (type and number of respondents used in field-testing, who, how and when items were written and reliability).

References

1. Wright, B. D., & Stone, M. H. (1979). *Best test design.* Chicago: Mesa Press.
2. Wright, B. D., & Masters, G. N. (1982). *Rating scale analysis.* Chicago: Mesa Press.
3. Wilson, M. (2005). *Constructing measures: An item response modeling approach.* Hillsdale, NJ: Erlbaum.
4. Bond, T. G., & Fox, C. M. (2001). *Applying the Rasch model fundamental measurement in the human sciences.* Hillsdale, NJ: Erlbaum.
5. Rasch, G. (1980). *Probabilistic models for some intelligence and attainment tests.* Chicago: University of Chicago Press. (Original work published 1960, Copenhagen: Danmarks Paedagogiske Institut.)
6. Byrk, A., Thum, Y., Easton, J., & Luppescu, S. (1998). *Academic productivity of Chicago Public elementary schools: A technical report.* Chicago: Consortium on Chicago School Research.
7. *Winsteps* (Winsteps download includes Ministeps). [Rasch Measurement Analysis Software]. Retrieved from http://www.winsteps.com/rasch.htm.

Chapter 11

Drawing Meaningful Conclusions from Education Experiments

Melanie M. Cooper

Department of Chemistry, Clemson University, Clemson, SC 29634

The ultimate goal of a chemistry education research project is that the scholarly work resulting from the research will have an impact on our understanding of teaching and learning, and will result in more effective and meaningful learning in the chemistry classroom. To do this the researcher must draw meaningful conclusions from the data, situate them within the context of previous work, and discuss their implications. It is particularly important that the researcher be aware of the kinds of errors that can creep into educational research, and this chapter describes a number of common problems often found in research reports in chemistry education. These include mistaking cause and effect, overgeneralization, using anecdotal evidence, not controlling for differences in student population, mistaking self-reported learning for actual improvements in student learning, and disturbing the test population by the act of investigation. Examples of exemplary reports using qualitative, quantitative, and mixed method research designs are provided to show how these problems can be avoided.

The preceding chapters in this book have discussed the major elements required for exemplary chemistry education research. Examples include: Zare (*1*) asked important questions in Chapter 2 that can be answered by research. Bunce (*2*) explained how to ask good questions in Chapter 4, which leads to designing appropriate experiments to approach the research questions, and using established techniques to analyze the data. As explained by Abraham and Williamson in Chapters 5 and 6 (*3,4*), the design of any experiment must be guided by theory. That is, there must be an underlying theory-base to any research, and any "treatment" administered must be based on that theory. Experimental designs may range from purely quantitative, as described by Sanger (*5*) in Chapter 8, where the data can be analyzed to produce statistically valid answers about what students learn, to qualitative designs (described by Bretz (*6*) in Chapter 7), where researchers directly study student behavior to try to understand why and how students learn. Mixed methods experiments that include both qualitative and quantitative techniques, described by Towns (*7*) in Chapter 9, can provide insights into both of these questions. Even if a research study has been well designed – with a question that can be answered by the methods chosen, and the data then analyzed correctly, there still remain a number of pitfalls for the researcher in moving to the next step of making the work useful to others by drawing meaningful conclusions.

This chapter focuses on how difficult it can be to drawing meaningful results from data. For example, researchers sometimes omit the last part of the scholarly process. They fail to ensure that the findings are explained and placed within the context of what is already known, and they fail to provide guidance and future directions. It has been said that "Research is a process for obtaining information, and scholarship is a process for converting information into knowledge."(*8*) For a research study to be a scholarly work, not only must it be situated in what is already known and be original and creative, it must also produce results that are meaningful and expand our knowledge of teaching and learning. This is not to say that a study is a failure if, for example, no significant difference in control and treatment groups is observed. As in any research study, a negative result can still have profound implications for future work in the area. Yet it is those implications that must be carefully presented and discussed for the research to become a work of scholarship.

Another aspect of producing meaningful results is avoidance of common mistakes that researchers may make in their desire to produce a finding that seems, on the surface, justified. This chapter will give a brief overview of some of the more commonly encountered mistakes.

Causality

Causality implies some dependent relationship between cause and effect. That is, if the cause occurs – then so does the effect. Educators in their eagerness to ascribe increased success to an educational intervention may sometimes forget the cardinal rule, that of examining and dealing with other confounding factors that must be ruled out or controlled. An example of this error is as follows (9). Over the past 200 years, the number of pirates has fallen in what appears to be a direct relationship to the increase in average global temperature. The conclusion that the loss of pirates has caused global warming or vice versa is clearly ludicrous. Obviously the two factors of pirate decline and global warming are in no-way directly related, yet this fundamental error is not so easy to discern when one of the variables **might** affect the other. This is especially true if the researcher starts with the prediction that one variable **will** affect the other. There are numerous reports in the literature of less egregious but still questionable conclusions, just as there are numerous reports indicating that some recent innovation in the classroom has affected student learning. Yet the authors have often neglected to address other factors that may be equally responsible for the observed effect. For example, a report (10) in the later section of this chapter on the question "Does going to Supplementary Instruction (SI) improve course GPA?" appears to be a relatively simple study. SI is one of many attempts to provide out of class support and guidance for problem solving to students in a broad range of courses.

At first glance it might appear that the data to collect would be the number of attendances at SI and the course grade earned by each student. Indeed there are a number of published studies that have found a positive correlation between attendance at problem solving sessions and course grade. Many of these studies suffer from a flaw, however, in that the student populations in the treatment and control groups are not properly controlled. A counter-claim to the idea that extra problem solving sessions improve overall course grades, might very well be that only the best, most diligent students actually attend these sessions, or that time on task produces gains in exam scores, not just time spent in SI sessions. Therefore the apparent correlation between attendance and achievement may not be valid, and the improvement in student grades could be due to student ability and perseverance. The problem with a simple correlation study is that the students who received the treatment (SI) were self-selected rather than comprising a random sample. It may be that the factors that cause students to attend extra problem solving sessions are also those factors that result in improved course grades.

Population Selection/Self Selection

The nature of chemical education research often dictates that we must work with the students and course organization that we have. Unfortunately this may not be the optimal research design. Many of the studies on interventions designed to improve the learning process do not have the option of using randomized control and treatment groups of students. While there are methods to factor out differences between control and treatment groups, as previously explained by Sanger (5), this step is sometimes neglected. If the student populations are self-selected, or teacher-selected, this may lead to problems with generalization of the results. A study of the sort discussed in the previous section, in which a self-selected group of students attends extra help sessions, is not persuasive unless the effects of self-selection can be factored out. Similarly, some studies on the positive effects of undergraduate research suffer from the problem that students are most often chosen for the study by the faculty, rather than being randomly picked from the available population.

Hawthorn and Pygmalion Effects

The Hawthorn effect is the name given to the proposition that the results of studies based on the behavior of a group of people may be biased if those people know they are being studied. It was originally proposed by researchers investigating worker productivity at a Western Electricity company plant in the 1920s and 30s. It appeared that productivity increased regardless of what change was instigated (lighting, pay, rests, etc.), and increased even when a change back to the original conditions was made. One common interpretation of these studies is that when participants know they are being studied, they may affect the outcome, but not for the reasons that were originally proposed. That is, the very act of observation changes the outcome of the study. Although still widely invoked, the original studies have been refuted (11) because of poor experimental procedures, but the related Pygmalion (12) effect has been validated in numerous reports. The Pygmalion effect could be considered as an educational placebo effect; students who are told that they are special and better than other students will perform at a higher rate than their peers. Conversely, as has been reported by Steele (13) and others, under-represented students who are asked to state their ethnicity on a test may score lower than they would if not asked for this information. In another study, women scored 15-30 percent lower than men on the Vandenberg test of mental rotations when subtly cued to think about their gender (14). If prompted to think about their exclusive status (for example as students in a private college), there was no statistical difference between the women's and men's scores.

Clearly the study of students' achievements and attitudes is rife with pitfalls; observed effects may, or may not, derive directly from the intervention. Just like in quantum mechanics when the observation of a system will perturb the system under observation, so it may be with people. If the subjects of the research are aware that they are being studied – and they should be aware since they should have signed Institutional Review Board forms giving informed consent – then the very act of studying their behavior may result in a different outcome than would have otherwise occurred. It is up to the researcher to design the experiment in such a way that these effects are minimized, can be factored out, or at least can be recognized.

Anecdotal Evidence

As observed by Rogers (*15*), faculty who develop educational strategies often promote them largely by personal experience. "I did this and my students liked it" is an alluring trap for some investigators. These faculty have neglected the basic tenets of research in their desire to communicate a new method or intervention. Certainly instructor enthusiasm can positively affect outcomes for students. But unless an innovation being investigated can be researched using the methods discussed in this volume, any report on the intervention cannot be meaningful to a wider audience and, as such, cannot be considered as scholarship. It is worth restating that evaluation has to be objective and reproducible. Anecdotal evidence is not research, particularly if it is reported by itself and collected in a haphazard fashion.

Use of Non-Validated Instruments

One of the most prevalent experimental designs is a pre-test / post-test design, in which a test or survey is administered to the population both before and after the intervention. While this can be a meaningful and reliable measure of the effectiveness of the intervention, many researchers use tests and surveys that they have constructed themselves without regard to the validity or reliability of the instrument. There are a large number of resources (*16*) to aid chemical education researchers in the construction of tests and surveys, and a brief overview of the field is given by Scantlebury and Boone in Chapter 10 (*17*). There are also validated surveys and tests available to the researcher, for example ACS exams (*18*) are useful in this context, and a variety of surveys such as the SALG (Student assessment of learning gains) can be obtained readily (*19*). What is important in the context of this discussion is the need to use tests and surveys that give reliable information. Most mid-term tests and exams have not been constructed to these rather exacting requirements, and so the results from the tests may not be meaningful.

Self-reported learning

It is not uncommon to find a paper in which the authors report the results of surveys in which students state that they learned more because of some teaching method or intervention. From these self reported surveys, the inference is then made that the students are correct. The fact that students happen to **think** they have improved their understanding and knowledge may be evidence of student satisfaction and a change in the affective domain, but it is obviously not in itself evidence of increased learning. Even when well developed and respected instruments for measuring student self-reported learning gains (the SALG, Student Assessment of Learning Gains) (*19*) and student satisfaction are used, they cannot be used alone as evidence of student improvement in comprehension or knowledge. Student attitudes and student learning are two different variables; and while one may affect the other, measuring attitude does not necessarily imply anything about learning.

Overgeneralization

Another common claim found in the literature is that the results of a particular study can be generalized to a quite different population. Undergraduate research is almost universally reported as being beneficial to a wide range of students, and there are numerous reports extolling the benefits of such experiences. However there is little evidence in fact that this is the case for the general population: most of the studies on undergraduate research are done with a highly selective cadre of students who are usually picked because of their aptitude in the field.

Such overgeneralizations may lead to the problematic conclusions that sometimes can occur when findings are extended from small to large enrollment courses. Many of the newer teaching strategies involve student groups, active learning, and inquiry. Yet most of the initial work on their effectiveness has come from studies in small classes taught by instructors who believe in the strategies, and who have the time and inclination to ensure that the new teaching methods "work". As many have found to their cost, the scale-up from small classroom to large lecture section requires more than the preparation of larger numbers of class activities. A system that produces measurable increases in learning and understanding in a small classroom, taught by a knowledgeable instructor, may fail completely when transferred to another institution. For example, instructors (or teaching assistants) who do not understand or buy-into the innovation can de-rail any anticipated progress. There are, of course, examples where research-based teaching methods have been incorporated successfully into large enrollment courses (*20*), but typically the research establishing their effectiveness was performed within the same type of environment, and the teaching and learning strategies were developed with the problems of large enrollment courses in mind.

Other problems may stem from extending the findings for students in one developmental level to students in a different level. Students who have not reached formal levels of logic, for example, may need different instructional strategies than those who have reached a higher level. Students who are at the "concrete" level of logical thinking do not benefit from being grouped together. This, in itself, is a good argument for not allowing students to form their own learning and study groups, especially since students often gravitate to others who are similar to themselves in ability. Many students, particularly ones at the lower levels of achievement, may not improve by practice alone. The strategy of "drill and kill" is not effective for these students and will not result in improvements in problem solving. Nor will it result in students who are interested and engaged in the subject (*21*).

Examples of Exemplary Papers

A Quantitative Study

The effect of supplementary instruction (SI) on students performances in chemistry (and other) courses has been the subject of numerous studies. A wide range of investigators have reported that this type of intervention produces improvements in student course performance, attitude, and retention not found for students who do not participate in SI. There is a fundamental flaw in most of these studies, however, in that students who participate in SI are self-selecting. In fact, this requirement that students not be forced to participate in SI is actually built in to the structure of the intervention. Therefore the students who do, and who do not, participate may be quite different populations and cannot be compared directly without further treatment of the data.

For example, one study (*10*) found a positive correlation between student attendance at SI and their standardized course grades ($r = .165$, $p < .001$). Yet these initial results were not persuasive because the student population was not controlled since SI is not mandatory and since other factors such as aptitude may have had a more powerful effect on course grades. Rather than merely reporting the simple correlation between course grades and SI attendance, these researchers also examined the relationship between the predicted GPA (PGPA) (a composite of high school GPA and ACT scores) and course grade, and PGPA and SI participation level. They found that PGPA was a strong predictor of course grade, but that no significant relationship existed between PGPA and SI attendance, indicating that there was no evidence that students with high PGPAs attended SI more. This supported the researchers' hypothesis that SI attendance has a positive influence on grades. The researchers also tried to control other variables, such as aptitude, class size, type of class (physical science or social science), by performing ANCOVAs (see analysis of covariance in Chapter 8 (*5*)) using the student's grade in the course under study as the dependant variable.

Using these methods the researchers were able to demonstrate that students who attended SI more than three times performed, on average, over half a grade point higher than students who did not attend SI after controlling for PGPA, student aptitude, student major, hours worked, or hours of planned study. These authors were able to go beyond their initial results by incorporating other information *ex-post facto* to produce much more compelling evidence for the effect of Supplementary Instruction. It is far more difficult to dismiss the results of a study of this type that has taken into consideration numbers of variables, than a study that looks only at the intervention and its effect on a selected group of students.

A Qualitative Study

A report by Seymour et. al. (22) approaches the evaluation of the benefits of undergraduate research using a long-term ethnographic study. As the authors discuss, there are numerous reports in the literature on the benefits of undergraduate research. They found nine reports in which the hypothesized benefits were both claimed and well-supported. However, the authors found over thirty reports in which the hypothesized claims were either merely stated or claimed without justification and not adequately demonstrated. A number of reports indicated that undergraduate research produces students who are more likely to be critical thinkers or to "think like a scientist", yet no evidence was offered to back up these claims. Of these reports, nine were actually evaluation reports of undergraduate research programs, yet all of them were judged to suffer from problems such as small sample size, or the fact that students were self-selected, or recruited by faculty to participate, and therefore predisposed to the expected outcome.

In order to overcome these problems Seymour and co-workers began a multi-institutional, long-term investigation to address a number of fundamental questions about the benefits and costs of undergraduate research. The research design for the pilot program reported in this paper involved a cohort of 76 students at three different institutions who were interviewed in depth three different times: first during the research project, second just before graduation, and finally after graduation. They were questioned about the nature, value, and career consequences of their involvement in undergraduate research. Faculty (N=55) were also studied and interviewed. This long-range, well-developed, study reported a number of findings that confirmed previous studies and impressions about the benefits of undergraduate research. Yet a number of the more common claims found in earlier papers were not supported. As example: most students did not present papers at scholarly meetings or co-author publications. The most important benefit cited by students was the increase in communication skills. This result was probably quite surprising to many faculty mentors, who may be more likely to concentrate their efforts on the increase in scientific and technical skills and competencies. There was no evidence that undergraduate research affected the choices made by students after graduation, but rather that the experience had clarified the students' pre-existing expectations.

The findings reported in this study, while surprising in some aspects, are much more persuasive because the authors approached the evaluation systematically using recognized research techniques, and they avoided the common pitfalls of self-reported evaluations.

A Mixed-Methods Study

Another exemplary study with profound outcomes for teaching and learning in chemistry is the report by Tien, Roth and Kampmeier on the effects of Peer-Led Team Learning (PLTL) in an undergraduate organic chemistry course (23). The authors presented a long-term study of an organic course that was transformed from a traditional lecture-recitation mode to a lecture-PLTL method. PLTL involves TA-led recitation (in which students ask questions and TAs show students how to solve problems and also give mini-lectures) was changed to a peer (undergraduate)-led, interactive group problem solving session (24). The study used a mixed methods design. Quantitative techniques were used to investigate relationships between PLTL and student performance and attitudes, while qualitative techniques, involving interviews, surveys, and journals of both students and PLTL leaders provided deeper insights into the reasons why changes in performance and attitude were observed.

The study compared two different populations of students: those enrolled in the course from 1992-1994 (recitation/traditional) and those from 1996-1998 ((PLTL). An examination of the two student cohorts revealed that they were not identical. The students in the PLTL cohort had significantly higher SAT scores than the 1992-4 cohort. It was also known that SAT scores are significant predictors of performance in this course. Since the researchers were aware of the differences between the two groups, they could use SAT scores as covariates to factor out group differences. In this way, by collecting more data than seemingly needed, and by using statistical techniques to correct for differences in populations, the researchers were able to provide some meaningful results without the need for a control and treatment group. While it is certainly easier to produce understandable and compelling results when there are two identical cohorts of students who treated in exactly the same way, it is often impossible to create this kind of ideal situation in the "real world".

This study found that students in the PLTL cohort significantly outperformed their counterparts from earlier years on all the course exams and obtained higher final course grades, even when the results were adjusted for SAT score differences. This is in itself a very important finding. Student improvement in scores was replicated over all the years of the study even as new peer leaders were trained and incorporated into the course each year. Students in the PLTL cohort also had higher retention rates. Yet this type of analysis gives no insight into why the use of peer-led workshops improved student performance. The researchers clearly had hypotheses about the reasons for improvement, since the PLTL movement itself is grounded in appropriate educational theory, including constructivism (25) and social development theory

(*26*). The researchers went on to investigate the reasons for improvement using a qualitative design.

Insights into the reasons for such improvements were obtained by observing, recording, and interviewing both students and peer leaders. Three common themes became apparent from these investigations.

1. Both students and peer leaders emphasized the idea of a workshop as a community of learners.
2. The workshop required students to self explain and negotiate among the group the meaning of what they were doing.
3. The workshops brought out the need to acquire expert thinking skills.

All of these themes are consistent with previously published research and ideas about why students learn so well in small group situations.

A third area of research reported in the paper brings us to the problematic area of self-reported learning and student satisfaction. Initial comparisons between the traditional and PLTL cohort showed no differences in student attitudes about PLTL and recitation sessions and their leaders. In fact the only differences to emerge were for questions that explicitly asked about interactions or actions that would most likely be associated with PLTL groups. For example, students reported that they interacted more with each other and understood how to work as a team if they were in the PLTL cohort. This part of the report shows how problematic it can be to rely only on students perceptions of what they learn, and how important it is to have other measures of student learning. The affective domain is, of course, a significant contributor to the learning process, and students who are dissatisfied with the learning environment are less likely to acquire the skills they need, but research that relies only on student attitudes about learning without actually investigating whether perceived improvements can be measured is not sufficient.

Conclusions

For educational innovations to take hold in the chemistry community they must be backed up by research that can be generalized to other situations and that produces meaningful results such as improvements in student learning, attitudes, and retention. Measuring and documenting these improvements can be more difficult than research in traditional chemistry arenas because of the wide range of confounding factors, the difficulty of working with human subjects, and the common mistakes that researchers make. That considered, convincing conclusions can be reached, and because the conclusions **are** convincing, they will have a profound and far-reaching effect.

References

1. Zare, R.N. In *Nuts and Bolts of Chemical Education Research*; D. Bunce and R. Cole (Eds.); American Chemical Society Symposium Series, Washington, D.C.: 2007.
2. Bunce, D.M. In *Nuts and Bolts of Chemical Education Research*; D. Bunce and R. Cole (Eds.); American Chemical Society Symposium Series, Washington, D.C.: 2007.
3. Abraham, M.R. In *Nuts and Bolts of Chemical Education Research*; D. Bunce and R. Cole (Eds.); American Chemical Society Symposium Series, Washington, D.C.: 2007.
4. Williamson, V. In *Nuts and Bolts of Chemical Education Research*; D. Bunce and R. Cole (Eds.); American Chemical Society Symposium Series, Washington, D.C.: 2007.
5. Sanger, M.J. In *Nuts and Bolts of Chemical Education Research*; D. Bunce and R. Cole (Eds.); American Chemical Society Symposium Series, Washington, D.C.: 2007.
6. Bretz, S.L. In *Nuts and Bolts of Chemical Education Research*; D. Bunce and R. Cole (Eds.); American Chemical Society Symposium Series, Washington, D.C.: 2007.
7. Towns, M.H. In *Nuts and Bolts of Chemical Education Research*; D. Bunce and R. Cole (Eds.); American Chemical Society Symposium Series, Washington, D.C.: 2007.
8. Assessing the Value of Research in the Chemical Sciences, National Academies Press, 1998, p. 86
9. Henderson, B. http://www.venganza.org/ (accessed September 4, 2006)
10. Kochenour, O.; Jolley, D.S.; Kaup, J.G.; Patrick, D.L.; Roach, K.D.; Wenzler, L.A. *Journal of College Student Development*, **1997**, *38*, No. 6
11. Franke, R.H.; Kaul, J.D. *American Sociological Review*, **1978**, *43*, 623-643
12. Rosenthal, R.; Jacobson, L. Pygmalion in the classroom: Teacher expectation and pupils' intellectual development. Irvington publishers: New York. 1992
13. Aronson, J.; Lustina, M.J.; Good, C.; Keough, K.; Steele, C.M.; Brown, J. *Journal of Experimental and Social Psychology*, **1999**, *35*, 29-46.
14. McGlone, M.S.; Aronson, J. Journal of Applied Developmental Psychology, **2006**, (accessed as uncorrected proof from science direct October 30, 2006)
15. Rogers, E. *Diffusion of innovations.* 4th ed. The Free Press. NewYork, NY, 1995
16. Thorndike, R M. Measurement and Evaluation in Psychology and Education: 7th Ed, Pearson Merrill Prentice Hall, 2005.
17. Scantlebury, K. and Boone, W. In *Nuts and Bolts of Chemical Education Research*; D. Bunce and R. Cole (Eds.); American Chemical Society Symposium Series, Washington, D.C.: 2007.

18. Division of Chemical Education – Exams Institute
    http://www4.uwm.edu/chemexams/ (accessed November 24, 2006)
19. Seymour, E. Student Assessment of Learning Gains.
    http://www.wcer.wisc.edu/salgains/instructor/ (Accessed September 5, 2006)
20. Oliver-Hoyo, M. T.; Allen, D.; Hunt, W. F.; Hutson, J.; Pitts. *J. Chem. Educ.* **2004**, *81*, 441.
21. Taconis, R. M.G.M. Ferguson-Hessler, H. Broekkamp ; *J. Res. Sci. Teach.*, **2001**, 38, 442.
22. Seymour, E:, Hunter A, Laursen S.L., Deantoni T. *Science Education*, **2004**, *88*, 493-534,
23. Tien, L.; Roth, Kampmeier, J. *J. Res.Sci. Teach.*, **2002**, *39*, 606-632
24. Gosser, D.K.; Cracolice, M.S.; Kampmeier, J.A.; Roth, V.; Strozak, V.S.; Varma-Nelson, P.(Eds.). *Peer-led team learning: A guidebook*. Prentice-Hall, Upper Saddle River, NJ 2001.
25. Bruner, J. *Acts of Meaning*. Cambridge, MA: Harvard University Press 1990.
26. Vygotsky, L.S. In *Mind in society: The development of higher psychological processes* (Cole M., John-Steiner V., Scriber S., & Souberman E., Eds. and Trans.). Cambridge, MA: Harvard University Press. 1978.

Chapter 12

Assessment of Student Learning: Guidance for Instructors

Christopher F. Bauer[1], Renée S. Cole[2], and Mark F. Walter[3]

[1]Department of Chemistry, University of New Hampshire, Durham, NH 03824
[2]Department of Chemistry and Physics, University of Central Missouri, Warrensburg, MO 64093
[3]Division of Sciences and Allied Health, Oakton Community College, Des Plaines, IL 60016

> Chemical educators are being challenged to think about student assessments that move beyond typical course exam scores. This chapter addresses how to set assessment goals, develop strong guiding questions, select appropriate tools and procedures, and collect and analyze data. Types of student learning discussed include knowledge, metacognition, attitudes, interactivity, communication, decision-making skills, and practical laboratory skills. Citations to well-documented instruments used to assess students in chemistry instructional settings are included.

"What are my students taking away from my course and their time with me?" Good instructors implicitly reflect on this question and, as a result, make adjustments to their classroom approaches. In order for teaching to develop as a scholarly activity and for student learning to advance beyond the status quo, it is important to turn this private reflection into an explicit, evidence-based process. This will help guide curriculum reform efforts, help support chemical education research explorations, and provide justification for instructional grants submitted to funding agencies (*1,2*). In part, this concern for evidence is coincident with recent and somewhat contentious discussions about what constitutes rigorous educational research (*3,4*) and about the validity and quality of large-scale, as well as classroom-scale, assessments (*5*). More pertinent, perhaps, is that many instructors of chemistry have been seeking additional means for assessing student learning to complement what course exams provide.

This chapter is *not* a comprehensive review of assessment instruments and approaches, nor is it a complete tutorial on assessment design and analysis (*6*). Other chapters in this monograph expand on those concerns and provide additional references and examples. The goal of this chapter is more simply to assist chemistry instructors in thinking about assessment in richer and more deliberate ways. Attention is focused on student outcomes, but much of the discussion and some of the instruments are applicable to a wider range of interests, including outcome assessment of programs and professional development of instructors.

Planning Effective Student Assessment

The following questions, discussed in this chapter, organize and help in planning effective student assessment:

1. What are the goals for assessment? What do I hope to learn and about whom?
2. How is the information to be used?
3. What questions can I ask? Are some questions better than others?
4. What have previous researchers learned about this question, and from what theoretical frameworks were they working?
5. How should I design data collection to obtain clear insights?
6. What data should I collect to answer each question? What would "good" data be?
7. How might I analyze the data?

In addition, this chapter will review the permissions needed to conduct data collection on human subjects, and whether one should use an existing assessment instrument or create one's own.

What Are the Goals for Assessment? What Do I Hope to Learn?

When chemistry instructors discuss assessment, most often they are talking about assessment of chemistry content knowledge – the traditional examination as exemplified by those available from the Examinations Institute of the American Chemical Society (*7*). There is a growing interest in achieving a broader set of student outcomes than just content knowledge, as described in the National Science Education Standards for K-12 (*8*) and in the newly conceived ACS Committee on Professional Training curriculum guidelines for chemistry degree programs (*9*). Table I lists a broad array of learning goals that one might intend for one's students. Careful thinking about student learning goals and about what one would like to learn from assessment will pay off when attempting to draw clear inferences from the results.

Table I. Learning Goals for Students

Learning outcome	*Description*
Knowledge	understanding of chemical concepts
Metacognition	ability to monitor and self-regulate mental efforts
Attitudes	thoughts and feelings about course experiences
Interactivity	ability to work as part of a group
Communication	ability to present information
Decision-making	ability to analyze and act on a laboratory challenge or problem scenario

How Is the Information to Be Used?

Student assessments can be classified as formative or summative. In terms of the single learning goal of content knowledge, formative assessment is intended to monitor student understanding while a course is progressing and to provide feedback that hopefully leads to improvement. For example, regular written or computer-based homework, periodic exams or quizzes, or project reports may provide the means for this feedback loop. In contrast, summative assessment is intended to specify a student's success in reaching the ultimate content goals of the course. For example, a comprehensive final exam assesses a student's understanding of the whole body of knowledge at the end of the course. The exams published by the American Chemical Society Examinations Institute are good examples of summative assessments (*10*). At many institutions, an evaluation of the student's performance is then made by combining the outcomes on all of these assessments into a single quality indicator, such as a letter grade.

Some of the non-content learning goals cited in Table I may be included in the overall evaluation (and grade) for a student. Metacognitive skills and decision-making ability may be assessed indirectly by providing novel or ill-defined exam problems or projects. Students who are able to apply their content knowledge more reliably or with more sophistication would be rewarded in the overall score. Interactivity may be assessed by means of rubrics completed by peers or by instructors to reward productive collaborative behaviors. Communication in written and oral expression may be evaluated in a similar fashion. These non-content learning goals would be appropriate to include if the curriculum and instruction helped students develop these abilities and the course syllabus was explicit about their importance. The only non-content learning goal that would be questionable to include as a component of a student's grade is attitude.

Assessment results for any of the learning goals are valuable for understanding how students engage with the content and procedures in a course, and for directing curricular improvements. Tables II to VI that follow contain specific references to assessment tools, research studies, or applications that illustrate diverse approaches to individual student assessment or to assessment of curricular success. For the most part, the references are drawn from the domain of chemistry.

What Questions?

Questions regarding each of the learning goals listed in Table I may be answered more definitively if they are specific and if they can be evaluated with reliable analytic methods. For example, the first question under *Knowledge Outcome* that one might ask is "Do my students know more chemistry as a result of a new instructional approach that I have taken?" Although this question embodies an important issue, it does not define what is meant by "know" nor what aspect of the disciplinary area of chemistry is of concern. A more specific question is "Have student scores on an ACS exam changed significantly as a result of the new instructional approach?" This identifies a specific means for assessment of chemistry content knowledge that leads to data that may be analyzed in a systematic fashion. It is not the only means for assessing content knowledge, but it will provide insight into the *Knowledge Outcome*. Similarly, "Are my students better at problem solving?" could be improved to "Have student abilities to apply what they know to problems that they have never seen before changed significantly?" The latter could be assessed through listening to a representative subset of students think through a novel problem out loud and analyzing their responses thematically (*11*). Tables II to VI, organized according to type of student outcome, suggest questions that one might ask initially (*common-language questions*) and questions that are refined and strengthened versions (*specific-language questions*), along with examples of the types of data or instruments one could use. These recommendations are suggestive rather than proscriptive. Leonard (*12*) provides a clear discussion regarding the asking of strong questions. In addition in this monograph, authors Zare (*13*), Bunce (*14*), Abraham (*15*), and Cooper (*16*) all discuss what questions are valuable and pertinent to investigate, how questions may be framed and refined, and what meaningful results look like.

Table II. Learning Goals for Knowledge

Common-Language Questions	Specific-Language Questions	Assessment Instruments and Examples
Do my students know more chemistry?	Have student scores on an ACS exam changed significantly?	ACS exam scores (*7*) Interviews (*17*)
Do my students perform better in the class?	Are student abilities to explain the reasoning behind a problem substantially different?	Exams that require explanation of reasoning (*18*)
Do my students solve problems better?	Have student abilities to apply what they know to problems (and problem types) that they have never seen before changed significantly?	Present student with a challenging problem. Record process through observation, audio, or video. (*19*) Use think-aloud protocol for individual students. (*11*)
Can my students think rationally?	To what extent does the reasoning ability of students develop?	Group Assessment of Logical Thinking (*20-22*), Test of Logical Thinking (*23*)

What Theoretical Frameworks and Prior Investigations Apply?

Over the past 30 years, there have been significant advances in understanding human learning and interaction such that a number of theoretical bases may be used to frame student assessment questions for all of the learning goals in Table I. This includes theories regarding higher order reasoning ability, personal epistemology, metacognitive ability, human and computer information processing, spatial acuity, mental schema, motivation, group dynamics, social and personal construction of knowledge, and affective (emotional or attitudinal) behavior. It is impossible to describe these thoroughly here. Readers are referred to Abraham in Chapter 5 (*15*) and to other introductory sources (*24-26*). In addition, Tables II to VI cite examples of theoretically-based assessments applicable in chemistry settings.

In a larger sense, assessment plans which build on the existing knowledge base and which are informed by pertinent theories of learning, are likely to provide more robust information (*5*). This information provides the context for deciding how to move forward, whether for an individual instructor interested in improving student outcomes at his/her own institution or for someone carrying out an education research investigation. Abraham in Chapter 5 (*15*) and Cooper in Chapter 11 (*16*) have argued that setting an investigation into intellectual context is a key characteristic of scholarly work. Williamson in Chapter 6 (*27*) gives a detailed example of how research regarding student understanding of the particulate nature of matter has informed curriculum, assessment, and further research.

How Should Data Collection Approach Be Designed?

If one's purpose is to study the effects of curriculum or instructional innovation on various student outcomes, an important issue in design is controlling threats to validity. Effects seen may not be due to the innovation but may be due to an uncontrolled factor. For example, in comparing Knowledge Outcomes one year to the next using scores from locally-constructed exams, it is important to ask for evidence regarding questions such as:

- Have exam questions been equivalent in coverage and challenge?
- Has grading from year to year used the same evaluation criteria?
- Has the content covered been equivalent?
- Have instructors employed similar instructional methods?

Any of these issues may become a source of bias or noise that could make it more difficult to see the effects of an intervention. Abraham and Cracolice (*28*) discuss this issue and how different experimental designs help control these factors.

Planning ahead is critical. It is not uncommon to encounter a situation in which curriculum changes are made for a current group of students to see how effective the changes are without clearly thinking about what comparisons should be made. Lack of forethought may compromise the curriculum intervention, the data collection, and the meaningfulness of the results.

The smaller the number of students, the stronger the need for qualitative data or longitudinal data. Frequently, educational experiments based on statistical inference lead to "no effect found." The reason could be that the

Table III. Learning Goals for Metacognition

Common-Language Questions	Specific-Language Questions	Assessment Instruments and Examples
Did students learn how to think clearly?	To what extent do metacognitive skills (such as self-regulation, elaboration, mental reorganization, and questioning) change?	Motivated Strategies for Learning Questionnaire (MSLQ)[1] (*29*) Interviews (*30*) Classroom observations: e.g. number or cognitive level of questions asked by students (*31*)
	How do cognitive expectations about learning chemistry evolve?	Cognitive Expectations for Learning Chemistry inventory (CHEMX) (*32*) Interviews
	How well do students evaluate each others' writing?	Calibrated Peer Review (*33*)

NOTE 1: MSLQ has six motivation subscales (intrinsic goal orientation, extrinsic goal orientation, task value, control of learning beliefs, self-efficacy for learning, test anxiety) and nine learning-strategy subscales (rehearsal, elaboration, organization, critical thinking, metacognitive self-regulation, time and study environment management, effort regulation.

Table IV. Learning Goals for Attitudes

Common-Language Questions	Specific Language Questions	Assessment Instruments and Examples
Have student attitudes improved?	Have student perceptions as learners of chemistry improved?	Chemistry Self-Concept Inventory (CSCI) (*34*)
		Motivated Strategies for Learning Questionnaire (MSLQ) (*29*)
		Student Assessment of their Learning Gains (SALG) (*35*)
		Chemistry Attitudes and Experiences Questionnaire (CAEQ) (*36*)
		Interviews
Do students think the course was worthwhile?	What are student perceptions of the relative utility and impact of course materials and techniques?	Student Assessment of Learning Gains (SALG) (*35*)
		Chemistry Attitudes and Experiences Questionnaire (CAEQ) (*36*)
		Interviews
Does this help students like chemistry class?	What percent of students withdraw?	Registration data from Institutional Research Office or Registrar
	What percent of students take more chemistry classes?	
	What percent of students switch to/from chemistry major?	

Table IV. *Continued.*

Common-Language Questions	Specific Language Questions	Assessment Instruments and Examples
Are students working hard?	How long do students stick with difficult problems?	Time on task persistence or patterns Electronic homework records Observations of individual student or student groups at work
Do my students have a better attitude towards chemistry?	To what extent have my students' attitudes changed regarding chemistry as a discipline?	Attitude-to-Chemistry Inventory (*37*) Chemistry Attitudes and Experiences Questionnaire (CAEQ) (*36*)
Are my students afraid of chemistry?	To what extent are students anxious about working in a chemistry laboratory?	Chemistry Laboratory Anxiety Instrument (CLAI) (*38*)

effect was small, that the human noise was large, or that the number of subjects was too small to overcome the noise. For the small college environment, attention would be more productively focused on collecting qualitative data about how students respond to the implementation and on establishing routines for collecting data longitudinally (over time) for a common course or for a common cohort of students moving through a sequence of courses. For larger institutions with multiple course sections, quantitative designs may be more immediately applied. Qualitative data will still be important for interpretation, as well as to aid planning long term comparisons.

What Data?

Many science instructors are interested exclusively in the *Knowledge* and *Skills Outcomes* and are familiar with typical student assessments such as tests and lab reports. It is often desirable to assess student learning goals that go beyond this into the affective and social realms. Collecting data pertinent to each of these goals requires different types of approaches or instruments. Although it is possible to construct one's own tools, there is a significant advantage (in time investment and quality of results) to using tools that are more rigorously evaluated. The particular examples cited here in Tables II to VI have undergone some degree of scrutiny regarding validity (i.e. that the results really indicate the mental construct or concept you think they are measuring) and reliability (i.e. that the results are reasonably precise and sensitive to student differences). Some of the tools are familiar content exams, some are surveys, some are based on observations or interviews, and some on written expression. It is important to note that the availability of rigorously-evaluated assessment instruments is limited (at this time) and that the examples provided must be evaluated for suitability to one's own situation. It is also important to note that a "rigorous" assessment doe not necessarily imply "numerical" or "quantitative" data. Important insights regarding student learning have resulted from studies that rely on descriptive qualitative studies involving observations, interviews, and samples of student work products (*39-41*). These important issues are discussed in detail in this monograph by Bretz (*42*), Towns (*43*), and Cooper (*16*).

Chemistry instructors consider the ACS Examinations Institute products as having substantial credibility (*7,10*). While these exams are an important resource for assessing the *Knowledge Outcome* for students, it is important to understand the limits of these tests. First, they are by design, summative – intended to ascertain student understanding at the end point of a course of study. Because of exam security, they are not suitable for providing student feedback. Secondly, they are broad. Major concepts are probed by only one or two questions. This makes ACS exams less sensitive for detecting changes in learning outcomes as a result of an innovation affecting only part of a curriculum. Thirdly, some exams focus more on recall of terminology and ideas, and application of procedural or mathematical skills. Other exams (labeled Conceptual, Special, or Paired Question) have been developed to tap different, conceptual aspects of content knowledge (*44*)

Table V. Learning Goals for Interactivity

Common-Language Questions	Specific Language Questions	Assessment Instruments and Examples
Has the course been effective?	How do students describe the effect the course has had on personal learning goals?	Learning Goals Inventory (*45*) Student writes reflective essay e.g. "Describe, using specific examples, how this course helped you meet your individual learning goals."
Are my students better at working in teams?	How do patterns of student interaction within groups change?	Video students in action Classroom observer to track participation rate, interaction patterns (*46*) Student reports
Do my students have better management skills?	Do students become more effective in working as a group to solve problems? (management skills, division of labor, communication)	Video Classroom observation of types of behaviors (task vs. process orientation) (*46*) Student reports

Collection of Data

Issues to consider regarding data collection may be categorized as intellectual and logistical. The former has to do with what one is trying to learn about individual students or about an instructional activity. The latter has to do with practical matters of how it will be accomplished.

Considering intellectual issues first, one may decide to collect data at a variety of times or from different sources depending on the student outcomes or contrasts of interest. Timing issues include pre/post comparisons and longitudinal studies. Population differences include comparisons between parallel student groups, or among subpopulations in a single classroom. These decisions will need to be made as part of the experimental design process. For example:

- Are students improving over time during this course?
- How do students experiencing instructional method A compare with students experiencing instructional method B?
- How do students this year compare with those from previous years?
- How do students perform in subsequent courses after their experiences in this course?

These questions all involve making comparisons. There are several issues to consider with designs of this type. First, the focus on instructional methods A vs. B can be interpreted too narrowly. For example, if students in group A (new approach) get exam scores that are the same as those in group B (old approach), is this evidence that the new instructional method "does not work?" Since an innovation may be designed to address more than test scores, the comparison would be richer if the question is asked differently: "In what ways do students in a particular classroom environment differ from students who have not experienced that environment?" This implies that multiple measures, particularly qualitative measures, should be used to elucidate what is happening. Another way to investigate outcomes is to ask "What types of students exhibit enhanced outcomes as a result of these experiences?" This allows for the possibility that students with different backgrounds or attitudes might respond differently. Thus, a whole-class average comparison may hide subtle effects for subpopulations of students. An example of this problem has been reported by Zusho, *et al*. (*47*) for college students in organic chemistry. Students' judgment of their confidence in doing well decreased over a semester. When split into a high versus low achievement groups, however, the high-achieving group gained in confidence, whereas the low-achieving group lost confidence.

The practical issues are concerned with how students are to provide information. If the information to be collected is embedded in student work (as discussed in the section on "How Is Information to Be Used?"), one need be concerned only with how to document these work products. If the information desired is external to the course (e.g. an attitude survey not included in course grade determination), the concern should be whether students are willing and able to respond adequately. Good information requires that survey respondents have enough time and desire to read, consider, and respond thoughtfully.

Inviting them inside the assessment process, to contribute their input as a means to improve the course for other students in the future, is a helpful introduction. One can also provide a small point bonus for their contribution.

Table VI. Learning Goals for Communication, Decision-Making and Laboratory Skills

Common-Language Questions	Specific Language Questions	Assessment Instruments and Examples
Can my students communicate better?	Have students improved their ability to communicate chemical information both verbally and in written form?	Interviews Quality of course products such as analysis of science article, position or research papers, presentation for designated audience, lab reports
Do my students handle themselves in lab better?	To what extent do students exhibit the ability to carry out particular laboratory tasks (e.g. proper use of balance, preparation of solutions)?	Observed lab behaviors evaluated with rubric (48) Performance assessment on special task (49)
Can my students solve lab problems?	With what facility are students able to work through decision-making for a laboratory problem?	Poster presentation of challenging task, experimental process, and results; peer and instructor ratings

How Should Data Be Analyzed?

The level of sophistication of data analysis will depend on the purpose for collecting the data—was the effort driven by a hypothesis regarding a curricular intervention, was it an exploration for potential patterns in survey data or themes in written student work, was it to obtain feedback regarding instruction, or was is to contribute to determining a student's grade. The following discussion is necessarily broad with appropriate references to more detailed resources.

The type of data in hand matters, whether quantitative (exam scores, survey scale responses) or qualitative (written student work, interview transcripts, observational records). The former are amenable to statistical analyses, *e.g.* *(50)*, Chapter 8 by Sanger *(51)*, Chapter 10 by Scantlebury and Boone *(52)*; the latter to qualitative or thematic analyses, *e.g.*, *(53)*, Chapter 7 by Bretz *(42)*, Chapter 9 by Towns *(43)*. In both cases, this is generally new territory for chemistry instructors. Human beings are much less uniform than molecules, and this added noise requires more sophistication in analysis in order to identify significant trends. To get the most out of the data one has, one must start to become informed regarding multivariate statistical techniques and the analysis of non-quantitative data. This would include study of texts, consultation with colleagues in the social sciences (education, sociology, economics, psychology), and advanced courses. There are also computer programs that can assist in data analysis. For statistical analysis, common software includes *SPSS (54)* and *SAS (55)*, the licenses for which are often held by one's institution. For qualitative documentation and analysis, one would use software such as *nVivo (56)* or *The Ethnograph (57)*.

For instance, assume the goal is to compare two different classroom environments. Let's also assume that the same instructor was involved in both, that exams were identical, and content coverage was similar. In addition, students were put into the two sections by the normal registration process with no prior knowledge of the type of classroom it would be. This is already a fairly ideal situation in which one could compare exam results. However, there is no guarantee of equivalence of student populations. One approach is to identify other possible variables that might affect performance and use those as controls (covariates) in the analysis (ANCOVA – analysis of covariance). Variables that have been used for this purpose include such things as math and/or verbal SAT scores, and performance in prior courses. Another approach is to select a matched set of students from the two classrooms who are similar on several characteristics (e.g. range of SAT scores, gender, year in school), except for the intervention under study.

Possible demographic characteristics that generally are in the institutional data base are gender, age, year in school, and major. Your colleagues in the social sciences or your office of institutional research likely have standard survey items to gather this type of information. Inquiring about other factors that have been correlated with school performance, such as ethnicity, socioeconomic status, or level of parental education is a bit more intrusive and needs oversight (see "Permissions" section below).

Possible performance indicators that could be used as covariates or for creating cohorts with specific characteristics are SAT or ACT scores, grade point average, or prior course grades. Doing a pre-test may also be helpful --this could include content tests such as the California Chemistry Diagnostic (ACS Examinations Institute) *(7)*, or cognitive instruments such as the GALT Group Assessment of Logical Thinking (a performance-based multiple-choice instrument based on Piagetian reasoning levels) *(20-22)*.

One important caution should be noted in use of inferential statistics. Performing a statistical test on two populations and finding "no difference" (e.g. a t-test on two means) is not the same thing as establishing a positive indication of equivalence between those two populations. This point is well argued in a recent article by Lewis and Lewis (*58*). This argument is also consistent with a growing concern among social scientists about the value of significance testing as compared with use of effect sizes or intervals (*59*).

What Permissions Are Needed?

If one is doing curriculum innovation and collecting data on students, this is human subject research that might legally and ethically require oversight by the institution's IRB (Institutional Review Board for the protection of human research subjects). The level of review depends on the level of risk for students. If the intent is to collect information from students for the purpose of improving a course and there is no intention of publishing the results, then IRB oversight is usually not required because this is what all good instructors do. In case of uncertainty, one's local IRB should be consulted. In the case of unusual time demands or risks for students, feedback from the IRB is strongly suggested.

If the intention is to collect information from students which *might* or *will* result in publication or presentation in a public or professional forum, then IRB oversight is often required. The IRB review assesses the risks and benefits for students in light of the purpose and nature of the study and determines what level of student consent is required. One issue that should not be forgotten is that faculty instructors are human subjects, too. If courses involving different instructors are to be compared, there are risks for the instructors and their consent is also required.

If IRB issues are new territory, the closest resources will be the IRB office and colleagues in psychology, sociology, and chemistry education research. The National Institutes of Health maintain a website and an on-line course (*60-61*). Local policies can vary.

Can I Design My Own Instruments?

Every chemistry instructor has written exams (the *Knowledge* learning outcome). Instructors who have ever prepared exams collaboratively with colleagues, such as in the ACS Examination Institute exam construction process (*11*), recognize the advantages of peer review, discussion, and field testing. The experience of having one's assessment questions reviewed by peers is humbling and should induce an appropriate level of caution and skepticism regarding individually-designed assessment instruments. Doing assessment well and in a way that serves a larger audience requires much more effort than is typically realized (see Scantlebury and Boone in Chapter 10 (*52*) and Cooper in Chapter 11 (*16*)). A chemical analysis analogy is appropriate. When faced with the need to measure some chemical parameter, is it better to choose a standard method that has been documented in the literature, or to build a method from scratch? A standard method has known performance characteristics (accuracy, precision, applicability to various matrices). Its limitations typically are understood. Results from various users in various applications can be compared. Instruments or procedures developed for student learning also benefit from investing effort in establishing validity and reliability. A locally-developed laboratory procedure or a locally-developed learning assessment simply has more risk associated with it. Tables II to VI cite a number of instruments and approaches that have been applied in chemistry instructional settings. Other sources for assessment that one might use directly or adapt may be found in the following citations (*62-65*).

Last Word

One last suggestion: find collaborators who have the expertise you need. If this were chemistry research, would it be productive to go it entirely alone when entering a new area? Probably not. Colleagues in chemical education research and in the social sciences are often happy to discuss issues that they face in doing their own research. The most important aspect of assessment of student learning is that it be done with as much care and deliberateness as one would exercise in a chemistry research investigation. Better decisions can then be made regarding student development, the effects of curricula and instruction, and program outcomes.

References

1. National Science Foundation. Course, Curriculum, and Laboratory Improvement (CCLI) Program Solicitation, NSF05-559. http://www.nsf.gov/pubs/2005/nsf05559/nsf05559.htm (accessed Fall 2006).
2. Bunce, D.; Gabel, D.; Herron, J. D.; Jones, L. Report of the Task Force on Chemical Education Research of the American Chemical Society Division of Chemical Education *J. Chem. Educ.* **1994**, *71*, 850.
3. *Scientific Research in Education*; Shavelson, R. J.; Towne, L., Eds.; National Academy Press: Washington, DC, 2002.
4. *Educational Researcher* **2002**, *31* (8), 4-29 (multiple authors and articles).

5. *Knowing What Students Know: The Science and Design of Educational Assessment*; Pellegrino, J. W.; Chudowsky, N.; Glaser, R., Eds.; National Academy Press: Washington, DC, 2001.
6. Bauer, C. F.; Cole, R. S.; Walter, M. F. POGIL Assessment Guide. http://www.pogil.org (accessed Fall 2006), The ideas in this chapter were initially developed as part of this on-line Assessment Guide.
7. American Chemical Society. ACS Examinations Institute Home Page. http://www3.uwm.edu/dept/chemexams/materials/exams.cfm (accessed Fall 2006).
8. National Science Education Standards, National Research Council, National Academy Press: Washington, DC, 1996.
9. American Chemical Society, Committee on Professional Training. http://www.acs.org (accessed Fall 2006).
10. Holme, T. *J. Chem. Educ.* **2003**, *80*, 594-596.
11. Bowen, C. W. *J. Chem. Educ.* **1994**, *71*, 184-190.
12. Leonard, W. H. *J. College Science Teaching*, **1993**, *23* (2), 76-78.
13. Zare, R. N In *Nuts and Bolts of Chemical Education Research*; Bunce, D. M., Cole, R., Eds.; American Chemical Society: Washington, DC, 2007.
14. Bunce, D. M. In *Nuts and Bolts of Chemical Education Research*; Bunce, D. M., Cole, R., Eds.; American Chemical Society: Washington, DC, 2007.
15. Abraham, M. R. In *Nuts and Bolts of Chemical Education Research*; Bunce, D. M., Cole, R., Eds.; American Chemical Society: Washington, DC, 2007.
16. Cooper, M. M. In *Nuts and Bolts of Chemical Education Research*; Bunce, D. M., Cole, R., Eds.; American Chemical Society: Washington, DC, 2007.
17. Teichert, M. A.; Stacy, A. M. *J. Res. Science Teaching*, **2002**, *39*, 464-496.
18. Cavallo, A. M. L.; McNeely, J. C.; Marek, E. A. *Int. J. Sci. Educ.* **2003**, *25*, 583-603.
19. Tien, L. T.; Teichert, M. A.; Rickey, D. *J. Chem. Educ.* in press 2006.
20. Roadrangka, V. *The Construction and Validation of the Group Assessment of Logical Thinking (GALT)*. E.D. Thesis, University of Georgia, Athens, GA, 1986. GALT© instrument available on request from Department of Science Education, University of Georgia (verified Summer 2005).
21. Roadrangka, V.; Yeany, R.H. *A Study of the Relationship Among Type and Quantity of Implementation of Science Teaching Strategy, Student Formal Reasoning Ability, and Student Engagement*; National Conference of Association for Education of Teachers in Science; Chicago, IL; Apr 1982; ERIC abstract ED215896.
22. Bunce, D.; Hutchinson, K. *J. Chem. Educ.* **1993**, *70*, 183-187.
23. Tobin, K.; Capie, W. *Educ. Psyc. Meas.* **1981**, *41*, 413-423.
24. McKeachie, W. J. *Teaching Tips*, 9th ed.; D. C. Heath & Company: Lexington, MA, 1994.
25. *Chemists' Guide to Effective Teaching*; Pienta, N. J.; Cooper, M. M.; Greenbowe, T. J., Eds.; Pearson Prentice-Hall: Upper Saddle River, NJ, 2005.

26. Herron, J. D. *The Chemistry Classroom: Formulas for Successful Teaching*; American Chemical Society: Washington, DC, 1996.
27. Williamson, V. M. In *Nuts and Bolts of Chemical Education Research*; Bunce, D. M., Cole, R., Eds.; American Chemical Society: Washington, DC, 2007.
28. Abraham, M.R.; Cracolice, M. S. *J. College Science Teaching* **1993/1994**, *23* (3), 150-153.
29. Pintrich, P.; Smith, D.; Garcia, T.: McKeachie, W. *A Manual for the Use of the Motivated Strategies for Learning Questionnaire. Technical Report 91-B-004*. The Regents of The University of Michigan, 1991.
30. Gopal, H.; Kleinsmidt, J.; Case, J.; Musonge, P. *Int. J. Sci. Educ.* **2004**, *26*, 1597-1620.
31. Pedrosa de Jesus, H.; Teixeira-Dias, J. J. C.; Watts, M. *Int. J. Sci. Educ.* **2003**, *25*, 1015-1034.
32. Bretz, S. L.; Grove, N. CHEMX: Assessing Cognitive Expectations for Learning Chemistry, 18th Biennial Conference on Chemical Education, Ames, Iowa, July 2004, Paper S329, http://www.users.muohio.edu/bretzsl/CHEMX.htm (accessed Fall 2006).
33. Calibrated Peer Review. http://cpr.molsci.ucla.edu (accessed Fall 2006).
34. Bauer, C. F. *J. Chem. Educ.* **2005**, *82*, 1864-1870.
35. Seymour, E.; Wiese, D. J.; Hunter, A.; Daffinrud, S. M. Creating a Better Mousetrap: On-line Student Assessment of their Learning Gains. ACS 218th National Meeting, entitled "Using Real-World Questions to Promote Active Learning", San Francisco, March 2000. http://www.wcer.wisc.edu/salgains/ftp/SALG PaperPresentationAtACS.pdf. Student Assessment of their Learning Gains (SALG) http://www.wcer.wisc.edu/salgains/instructor/ (accessed Fall 2006).
36. Dalgety, J.; Coll, R. K.; Jones, A. *J. Res. Science Teaching* **2003**, *40*, 649-668.
37. Bauer, C. F. Validation of Chemistry Self-Concept and Attitude-toward-Chemistry Inventories", ACS Annual Meeting, Paper #1172, New Orleans, March 2003.
38. Bowen, C. W. *Educ. Psych. Meas.* **1999**, *59* (1) 171-185.
39. Phelps, A. J. *J. Chem. Educ.* **1994**, *71*, 191-194.
40. Adamchik, C. F., Jr. *J. Chem. Educ.* **1996**, *73* (6) 528-531.
41. Oliver-Hoyo, M. T. *J. Chem. Educ.* **2003**, *80* (8) 899-903.
42. Bretz, S. L. In *Nuts and Bolts of Chemical Education Research*; Bunce, D. M., Cole, R., Eds.; American Chemical Society: Washington, DC, 2007.
43. Towns, M. H. In *Nuts and Bolts of Chemical Education Research*; Bunce, D. M., Cole, R., Eds.; American Chemical Society: Washington, DC, 2007.
44. Moore, J. W. *J. Chem. Educ.* **1997**, *74*, 477.
45. Walter, M. A Window Into Students' Learning Goals: The Learning Goals Inventory, AAHE National Meeting, Chicago, Mar 2002.

http://www.oakton.edu/user/~mwalter/lgi/LGI.htm and
http://www.oakton.edu/user/~mwalter/lgi/LGI_Poster_no_plots_3_12.htm (accessed Fall 2005).
46. Johnson, D.W.; Johnson, F. P. *Joining Together: Group Theory and Group Skills*, 8th ed.; Allyn & Bacon: Boston, MA, 2002.
47. Zusho, A.; Pintrich, P.R.; Coppola, B. *Int. J. Sci. Educ.* **2003**, *25* (9), 1081-1094.
48. Oliver-Hoyo, M. T. *J. Chem. Educ.* **2003**, *80*, 899-903.
49. Mills, P. A.; Sweeney, W. V.; DeMeo, S.; Marino, R.; Clarkson, S. *J. Chem. Educ.* **2000**, *77*, 1158-1161.
50. Lewis, S. E.; Lewis, J. E. *J. Chem. Educ.* **2005**, *82*, 135-139.
51. Sanger, M. J. In *Nuts and Bolts of Chemical Education Research*; Bunce, D. M., Cole, R., Eds.; American Chemical Society: Washington, DC, 2007.
52. Scantelbury, K.; Boone, W. J. In *Nuts and Bolts of Chemical Education Research*; Bunce, D. M., Cole, R., Eds.; American Chemical Society: Washington, DC, 2007.
53. Greenbowe, T. J.; Meltzer, D. E. *Int. J. Sci. Educ.* **2003**, *25* (7), 779-800.
54. SPSS Inc. SPSS Home Page. http://www.spss.com (accessed Spring 2007).
55. SAS Institute Inc. SAS Home Page. http://www.sas.com (accessed Spring 2007).
56. QSR International Pty Ltd. nVivo Home Page. http://www.qsrinternational.com/ (accessed Fall 2006).
57. Qualis Research Associates. Ethnograph Home Page. http://www.qualisresearch.com/default.htm (accessed Fall 2006).
58. Lewis, S. E.; Lewis, J. E. *J. Chem. Educ.* **2005**, *82*, 1408-1412.
59. Kline, R. B.; *Beyond Significance Testing: Reforming Data Analysis Methods in Behavioral Research*; American Psychological Association: Washington, D. C., 2004.
60. National Institutes of Health. Office of Human Subjects. http://ohsr.od.nih.gov/index.html Research (accessed Fall 2005).
61. National Cancer Institute. Cancer.gov—Human Participant Protections Education for Research Teams, http://cme.cancer.gov/clinicaltrials/learning/humanparticipant-protections.asp (accessed Fall 2006).
62. Angelo T. A.; Cross K. P. *Classroom Assessment Techniques: A Handbook for College Teachers*, 2nd ed.; Jossey-Bass: San Francisco, CA, 1993.
63. Practical Assessment, Research and Evaluation (PARE). http://pareonline.net/ (accessed Fall 2006).
64. Burros Institute of Mental Measurements. http://buros.unl.edu/buros/jsp/search.jsp (accessed Fall 2006).
65. Educational Testing Servcices. Educational Testing Service Test Collection. http://www.ets.org/testcoll/index.html (accessed Fall 2006).

Chapter 13

Collaborative Projects: Being the Chemical Education Resource

Barbara A. Sawrey

Department of Chemistry and Biochemistry, University of California at San Diego, La Jolla, CA 92093-0303

Chemistry education specialists in higher education should be considered as resources. One way in which chemistry departments can highlight and use such a human resource to great advantage is as part of a collaborative project. The chemical educator can contribute a scholarly education component to projects that could involve departmental colleagues, faculty from other departments or universities, community members, or K-12 school districts. Such collaborations can benefit everyone involved and have impact on science and education research. It is important that the chemical education component be meaningfully integrated into the larger project and be appropriately funded.

Introduction

Chemical education researchers are often the only such specialist in a chemistry department. This situation often leads to possibilities of collaborating in the research projects of other faculty who need an "expert" in chemical or science education research as part of their team. Chapter 3 (*1*) describes how a chemical education researcher can build collaborations to support his or her own research. This chapter discusses the many ways in which being the chemical education expert in a department or on a campus can provide additional opportunities to work with colleagues in science and engineering who carry out traditional research, with school of education colleagues, with K-12 systems, or in community partnerships. Regardless of the size of the collegiate institution, its location, or its specialization, there are opportunities to participate in collaborative projects that can impact science and education research.

Collaborative projects are a good way for a chemical educator to get started in the education research arena, and a way for the experienced chemical education researcher to lend expertise to projects that are enhanced by an education research component. The nature of the collaboration can take many forms, depending on the other members of the team, the possible funding source, and the research proposed. The chemical education researcher's role can vary widely, though no one can be an expert in every area. Some common opportunities are detailed in the rest of this chapter.

Collaborations

Working with Science and Engineering Faculty

Many research projects must include education, outreach, and evaluation components, even if the research is in a traditional science or engineering field. Federal agencies often require that proposals contain a plan for bringing the results of fundamental research to the classroom, teaching laboratories, or general public. Almost all agencies and foundations require evaluation of a research project. Here is where a chemical education researcher can play an important role (without being the lead principal investigator – the PI) as part of the team. Whether this is a project with a single scientist or engineer, a large interdisciplinary team, or is a training grant to fund graduate students and postdoctoral fellows, there can be a demand for an education piece. Science and engineering faculty members who carry out traditional research can be eager to work with a chemical education research specialist – someone who understands how to design and evaluate education research and curriculum development. This is a win-win situation, and a chance to influence projects by providing a scholarly educational component. It is important that the education component not be simply an add-on, but be integrated into the project in a logical and credible fashion. This is, of course, what the agencies expect. As an example, below is a quote from the program solicitation (2) for the National Science Foundation's cross-cutting Materials Research Science and Engineering Centers (MRSEC, a program sponsored by the Directorate for Mathematical and Physical Sciences, Division of Materials Research):

> ***Education, Human Resources Development.*** *Describe the education and human resource goals, provide a rationale for those goals, and indicate desired outcomes for the 6 year period. Briefly describe how the education goals integrate strategically with the research and organizational/partnership opportunities of the Center. Outline plans for increasing the participation of women and underrepresented minorities in Center research and education activities. Outline plans for seminar series, colloquial workshops, conferences, summer school and related activities, as appropriate. Describe any additional education programs not included in other sections of the proposal.*

The solicitation specifically calls for incorporating the educational plan into the overall program goals, with a special reference to including women and underrepresented minorities. This too is a recurring theme highlighted by funding agencies, and another reason that traditional researchers turn to education specialists for assistance. There is no simple resolution to an issue as complex as increasing diversity in the science and engineering pipeline (and one must be cautious not to promise more than can be delivered), but solid strategies for inclusion can be designed. For instance, multi-institutional projects can include institutions that have a diverse population, such as those serving predominantly underrepresented groups. Or if professional pipelines are being developed through the recruitment of students, the inclusion of community college students might be included in the proposal.

The laboratory researchers cannot operate separately from, even if in concert with, the education component. The NSF MRSEC solicitation refers to this directly.

> *One of the principal strategies in support of NSF's goals is to foster integration of research and education through the programs, projects, and activities it supports at academic and research institutions. These institutions provide abundant opportunities where individuals may concurrently assume responsibilities as researchers, educators, and*

students and where all can engage in joint efforts that infuse education with the excitement of discovery and enrich research through the diversity of learning perspectives.

In other words, the agency expects to see evidence that the research and education components blend together well. The proposal should articulate how the personnel will collaborate, and how the project will benefit from the synergy. For example, there are successful MRSEC centers that provide educational opportunities for students from underserved populations in their geographical locale, that have special summer programs for high school teachers, or that involve curriculum development directly related to the scientific research (*3*). All these situations provide excellent possibilities for the teaming of researchers and educators.

Training grants provide another opportunity for chemical education involvement. Typically training grants focus on the development of a special pathway for educating graduate students and/or postdoctoral researchers in emerging interdisciplinary fields, where faculty from multiple departments are working together, and where there is a need for organizing the cross-training of everyone involved. This may lead to new courses and seminars being set up, recruitment and retention of graduate assistants, professional development events for students, the training of faculty across seemingly disparate fields, and evaluation of the effectiveness of the program. Any one of these program elements can benefit from the planning, development, and assessment viewpoint of the chemical education researcher. For instance, a project might offer graduate students or postdoctoral researchers special opportunities to teach, maybe as preparation for an academic career. The chemical education specialist could work as a mentor, trainer, and/or evaluator of these new instructors.

Working with Schools of Education

When the chemical education researcher works with colleagues from the traditional science and engineering disciplines it is often because he/she is viewed as the *education specialist*. Whereas when the same researcher works with colleagues from a college or school of education, he/she is more likely to be viewed as the *science specialist*. The perspective of the chemical education researcher may or may not differ in these two instances, but the milieu differs, as do the expectations from the cooperating audience. The goals of projects of education school researchers and those of traditional science researchers are usually dissimilar, even in areas of outreach. As part of an education proposal the chemical education researcher may be asked to serve as a content expert, reviewing curriculum, but not to oversee an assessment plan. Whereas, as part of a science proposal the chemical education researcher may be asked to design and carry out an evaluation plan, but not get involved in the curricular content. Or, one may be asked to serve on an advisory board for the project, or even to lend support and name recognition as a Co-PI. It is important to define the role in advance that each team member will play in the project. Any successful collaborative proposal will explain why and how the chemical education researcher is involved, and how the project's impact is enhanced by this.

Working with K-12

Projects between colleges or universities and the K-12 sector, whether with an individual school or an entire district, require sensitive handling. It can be a challenge to maintain a partnership of equals. Many projects funded at the federal level require such cooperative ventures, however. The larger infrastructure, dependence on extramural funds, and higher indirect cost rates in post-secondary education often require that the university be the managing or senior partner in projects with K-12. The projects might involve pre-service teacher training, in-service teacher training (also known as professional development), science fair projects, student visits to campus (or the reverse – scientists visiting school classrooms), or curriculum development, among many other possibilities. By establishing strong ties to local schools, scientists and science educators put themselves in a position of being a resource to local teachers and administrators. This can open the door for involvement in some very large systemic projects, and with agencies and directorates not otherwise available for funding chemical education research. For example, many states have initiatives to improve the teaching of chemistry, or to provide in-service training for teachers using what are called Title I funds (4). (Title I of the federal Elementary and Secondary Education Act of 1965 is designed to help disadvantaged children meet challenging content and student performance standards. States receive federal funding that must be tied to Title I requirements.)

Some of the issues to consider when working collaboratively with the K-12 community are:

- Teachers have full days already, so any extra time needs to be compensated with a stipend, especially for evening, weekend, or summer activity.
- Professional development should also allow for teachers to earn continuing education credit from the university.
- Public K-12 teachers do not have the same curricular freedom as university faculty, and operate in a different political environment. They must meet the requirements of their school, district, and state board of education.

- Principals and sometimes district personnel should be made fully aware of your efforts and must approve a teacher's involvement.
- Special precautions and permissions are required to work with students under the age of 18. Institutional Review Boards must approve their involvement, and the parents of students must sign appropriate permission slips. (Refer to the section of this chapter titled "Institutional Review Board Approval" for more information.)

Community Partnerships

Community partnerships can involve multiple organizations that cut across the business-academic divide, and often include local philanthropic organizations. These partnerships are particularly attractive to regional business and government sectors, since they can have local economic impact. Many cities and states have foundations that focus their attention and money on education, though rarely on educational research. Nevertheless, these sources of funding for synergistic activities that can have a chemical education research focus should not be overlooked. There are many ways to find out about local philanthropic organizations, some of which are listed below.

- Ask a local fundraiser or development officer at the university for suggestions.
- Look at the publications of local museums to see what organizations help fund their events or activities.
- Online searching for "grantmakers" will yield links to lists of foundations and granting organizations that can be sorted by locale. A more specific example – searching on the words "philanthropy" and San Diego" for instance, turns up a list of granting agencies in the San Diego area that have websites.

Interinstitutional Projects

Projects that involve multiple institutions of any academic level, or informal learning organizations (e.g. museums) can introduce additional layers of complexity – even as they broaden the range of impact – when compared to working within one institution. It is necessary to identify one PI and a lead institution, though there may be a number of co-principal investigators. The institutions other than the lead may receive funding through subcontracts, or as a group of smaller proposals (one per site) bundled together in one package. Since each site will have its own rules for applying for extramural funds, and its own indirect costs rates, the flow of overall work and funding must be carefully negotiated and delineated. The local cultures and politics can make interinstitutional collaborations a challenging venture, but written agreement from the groups involved can provide reassurance that all will work smoothly.

All large projects need an Advisory Board, but it is especially important with interinstitutional projects that such a board include members from each participating establishment. This board can be used as a way of assuring the funding agency that the project will remain focused on its goal. The board can provide periodic (usually annual) reports to the funder on the impact of a program. The board members can also vet new ideas or modifications to the program, review evaluation plans, and help to disseminate successful features. An Advisory Board can be an important ally to the researchers, and provide valuable and timely feedback that strengthens the project.

Although the constituency of an Advisory Board should be discussed with all the critical personnel involved in a project, some examples of members who might specifically be included as relevant for the chemical education research piece of a project are shown below.

- Example 1: A joint project between a science department and a school of education will provide professional development for high school teachers in order to update and expand their science knowledge. In this case the board might include a school district representative.

- Example 2: A science research project in an emerging interdisciplinary field will also fund the development of new curricular material for a high school or college-level course. One of the advisors in this case might be a faculty colleague from sociology or psychology who can review the evaluation plan.

Theoretical Perspective

Regardless of the size of the role played by chemical education research in a collaborative project, it is important to identify the theoretical perspective of the education component. The theoretical perspective is the lens through which the education component is viewed. This should be explicitly stated. Any project needs goals and a framework against which outcomes can be measured. The theoretical perspective also directs the researcher toward certain methodologies that may be more appropriate than others. Chapter 5 (5) of this book includes a discussion of why this is important and explains a number of important theoretical frameworks.

Listed below are some of the questions that the chemical education researcher might ask when considering the theoretical framework and the design of the chemical education research component.

- Are there pre-existing characteristics or conditions to document (such as student achievement levels, or frequency of involvement in an activity)?
- What is the range in age and background of the participants? This is important since such factors are variables that may need to be factored into the analysis of results.
- Are there qualitative and/or quantitative components to the study? For instance, will test results be monitored? Will there only be numerical data collected (e.g. how many students answered a question correctly)? Will there be interviews conducted to probe student beliefs or thinking?
- Is the goal organizational or cultural change? In other words, is there the expectation of affective change, or a change in the infrastructure of a system?
- Is simple participation (of students or faculty) part of the goal?
- Will there be curriculum development involved? If so, that may need to align with state and local standards (in K-12), or with that acceptable to a curriculum committee (in a college setting).
- Which is more important – process or outcomes? Is it only a result that matters, or are there things to learn about how the result comes about?

It is critical to know what other team members have in mind for research, evaluation, and outreach, since the questions they want answered can affect the way the project is designed.

Institutional Review Board Approval

Whenever research involves collecting data about students (whether or not they are individually identified), or monitoring changes or outcomes in student behavior or performance, it is necessary to have approval to conduct this research from your campus Institutional Review Board (IRB) or Human Subjects Committee. This committee is charged with overseeing research that involves people. The IRB assures that human subjects are treated ethically, know their rights, and are alerted to the benefits as well as potential risks of involvement in a study. All surveys, interview questions, audio/videotaping requests, uses of data, longitudinal monitoring, photography, and so on, must have approval from the local IRB prior to acquisition. This process is governed by federal regulation, the details of which can be acquired from any college or university IRB, or online at any federal funding agency website. The National Cancer Institute offers a free online tutorial program (6) that education researchers may find helpful. Funding agencies typically require proof of IRB approval before awards are made to an institution or PI, although proposals can be submitted without having yet filed for IRB approval. Many journals also require proof of

IRB approval prior to publishing an article that contains student data. Each academic institution has its own method of reviewing requests to work with human subjects. Researchers should contact their IRB well in advance of collecting data in order to learn the submission process.

Interinstitutional projects may require IRB approval at multiple sites, each in a different format. It should be kept in mind that students younger than 18 cannot give consent to use information regarding their own data. It is their parents who have the authority to sign release forms and give informed consent. This can be a particularly time-consuming part of a project when working with K-12 students, and must include the school authorities. The principal must be aware of the project. The classroom teachers are also an ally whose cooperation is critical. It may be helpful to include funds to compensate the teacher for time spent collecting consent forms.

Budget

A "rule of thumb" for large projects requiring evaluation and assessment would be to allocate 10% of the overall funding for this purpose. Some federal agencies indicate this level of support is expected, while others are vague, so be sure to read the request for proposals carefully. The actual amount would vary depending on the nature of the work, but items to keep in mind when considering a budget are appropriate requests for:

- salaries and stipends (learn the difference between the two at your institution), projected with reasonable annual cost-of-living increases
- graduate assistant stipends, and perhaps hourly support for undergraduates participants
- participant costs
- travel
- supplies and expenses
- workshop costs
- equipment

Budget justifications must include an explanation of what each budget item is and why it is needed. Not all budget items are subject to recovery of indirect costs.

Letters of Support

Letters of support can be an important element of a collaborative proposal to a funding agency. The value of the collaboration can be emphasized through these letters, and the reviewers can gain insight into how the team members will work together. The letters can address how the whole is greater than the sum of parts, as well as the support that exists in a department, school, or university for the chemical education research component of a project. The role the person will play in the project should also be spelled out in the support letter. School districts' letters should mention previous collaborations with the PI's institution that have been fruitful. Administrators, such as a chair, dean, or vice president, can provide a more global perspective about how the campus views the proposal. Such a letter can indicate local support already in place to help the project, such as space, administrative assistance, etc. All named Advisory Board members should also submit a letter indicating their support and willingness to serve on the board. A template letter that they can modify can be provided to them by the principal investigator(s).

Benefits and Cautions

Many of the benefits to a chemical education researcher of being involved in a collaborative project are apparent. It is an opportunity to work with colleagues, to learn about the topic from their perspective, and share some of the questions that chemical education research can answer. Collaboration can provide intellectual and monetary support for projects, and help broaden the audience for outcomes. Being proactive in generating collaborations can lead to fruitful synergies, even without funding. Look for opportunities to foster joint projects or to either advise others or ask for their advice. Be broad in casting your net for possible connections.

The chemical education researcher must also be alert for the challenges that will appear, some of which are mentioned in other sections within this chapter. Every department and institution has its own culture of research and its own bureaucracy to be navigated. IRB approval may be necessary. By definition collaboration means that no one researcher has complete control over the research questions to be investigated or the methods used. The value of outcome and process may be viewed differently by the team members. As with any project, the budget may constrain what can be done. The vocabulary used by science education researchers can have unintended meanings when used with scientists and engineers, just as science vocabulary can have unintended meanings for education colleagues. So the chemical education researcher must pay particular attention to communication with faculty across different fields, and must clarify definitions of pertinent terms.

Most important of all the chemical education research contribution to a project must be an integral part of the overall project, rather than a second thought, and should have a clear, separate budget. Even though the partnership may be newly established for the purpose of a particular project, it will be obvious to reviewers and the agency if this is only "a collaboration of convenience". By developing relationships and networks prior to proposals, the plan is likely to be more cohesive, believable, and authentic.

Recommended Readings

Books

Isaac, Stephen and Michael, William, *Handbook in Research and Evaluation: A Collection of Principles, Methods, and Strategies Useful in the Planning, Design, and Evaluation of Studies in Education and the Behavioral Sciences*, Edits Pub., 1995.

Mertens, Donna M., *Research and Evaluation in Education and Psychology: Integrating Diversity with Quantitative, Qualitative, and Mixed Methods*, 2nd ed., Sage Publications, 2004.

Rossi, Peter H., Lipsey, Mark W., and Freeman, Howard E., *Evaluation: A Systematic Approach*, SAGE Pub., 1999.

Websites

National Science Foundation information regarding IRB requirements can be found at: *http://www.nsf.gov/bfa/dias/policy/docs/45cfr690.pdf*

Western Michigan University's Evaluation Center has a wealth of free information, checklists, and links to other sites: *http://www.wmich.edu/evalctr/*

References

1. Jones, Loretta L., Scharberg, Maureen A., and VandenPlas, Jessica R. in *Nuts and Bolts of Chemical Education Research*: D. Bunce and R. Cole (Eds.): American Chemical Society Symposium Series. Washington, D.C.: 2007.
2. Materials Research Science and Engineering Centers (MRSEC). National Science Foundation Program Solicitation. http://www.nsf.gov/pubs/2004/nsf04580/nsf04580.htm (January 2007)
3. Materials Research Science and Engineering Centers Website. http://www.mrsec.org/ (January 2007).
4. U.S. Department of Education. http://www.ed.gov/programs/titleiparta/index.html (January 2007)

5. Abraham, Michael R. in *Nuts and Bolts of Chemical Education Research*: D. Bunce and R. Cole (Eds.): American Chemical Society Symposium Series. Washington, D.C.: 2007.
6. National Cancel Institute. Human Participants Education for Research Teams. http://cme.cancer.gov/clinicaltrials/learning/humanparticipant-protections.asp (January 2007)

Chapter 14

Building a Fruitful Relationship between the Chemistry and Chemical Education Communities within a Department of Chemistry

Gabriela C. Weaver

Department of Chemistry, Purdue University, West Lafayette, IN 47906–2084

Many chemical education researchers are faculty members within departments of chemistry at higher education institutions. The type of research work that they carry out has fundamental similarities to other scientific research in the department, but differs in its methods and tools. Research in chemical education can be of great benefit to departments and other scientists in those departments. However, the work and the field may not be fully understood by all of the stakeholders. This chapter describes ways in which the goals of chemistry departments and chemical education researchers can be identified and overlap between them can be found. Descriptions of possible modes of collaboration and interaction are provided. The chapter concludes with a discussion of the importance of communication among chemical education researchers, other scientists and the broader public.

© 2008 American Chemical Society

This volume began with a call for improved communication between chemical education researchers and other chemists. One of the first places where this should occur is in the home institutions and departments of each chemical education researcher. In order for any researcher to be successful and to be fulfilled in his/her work, it is necessary for a good relationship to exist between the individual and his/her department. If good communication and mutual respect are achieved, the relationship can be a fruitful one for both the individual researcher and the department as a whole. This chapter will provide an overview of issues to consider and strategies to follow in establishing a fruitful relationship between a chemical education researcher and the rest of the chemistry community in that person's department.

Chemical education researchers may play various roles within departments of chemistry. The nature of the relationship will be determined by characteristics of the department – size, programmatic emphasis, history – as well as the background, expertise and interests of the researcher. In many cases, researchers in chemical education are seeking tenure through departments of chemistry that do not have a long (or sometimes any) history of tenure decisions in the field of chemical education.

Building a fruitful relationship from the beginning is in the best interest of both the department and the chemical education researcher. The beginning of such a fruitful relationship actually occurs *before* a chemical education researcher is hired. The faculty of a department must have both an understanding of and a commitment to the research field of chemical education, in much the same way that they have and understanding of and commitment to physical chemistry, organic chemistry or the other subdisciplines of the science. While developing an understanding of the field does not necessitate expertise in its theories and techniques, it does require that some level of consensus exist in the department about the expectations for the role of a researcher in chemical education. This is a critical foundation to be laid for the success of that person as a researcher and a colleague.

Similarly, the chemical education researcher must have an understanding of the priorities within a particular department. As is true for any researcher in a department, these priorities will influence and shape the work environment for the chemical education researcher. Any researcher in a chemistry department will need to give highest priority to developing as strong a research program as possible. However, due to its inherent overlap with the educational mission of a department, faculty colleagues and administrators in a department will have a variety of ideas about the role that an *educational* researcher should play within that department. During the hiring process, and early in the relationship, it is important that the assumptions and expectations be discussed and, when appropriate, modified so that a mutually beneficial research program can be established. Beginning from a point of mutual understanding, communication and cooperation are necessary for the research program of a chemical education researcher to be able to succeed within the context of a particular departmental environment. Ensuring that both the department and the researcher have similar goals in mind for that person's role in the department is critical to forming a fruitful relationship.

Characteristics of the Department

It is useful to examine the types of goals that departments of chemistry may have, and the various forces that shape and influence those goals. Departments that have a research mission will be concerned about issues related to the publication and grant writing success of their faculty. This, in turn, will influence priorities with respect to space, hiring and budgeting. The research mission may be a primary component of the department's financial sustainability. The degree to which this is the largest or a much smaller component is defined by the individual department's character and overall mission. The exact nature of the research mission will then have an impact on faculty in a given department in the form of expectations for research support of summer salary, various operating expenses of the research group or the department and, in some cases, support of graduate research assistants.

The department will have an educational mission as well. Again, this will vary in emphasis depending on the type and size of the institution and department. The mission will depend on the audience that the department and the institution serve. The educational background, degree objectives, geographical distribution and demographics of the student population will all serve to shape the educational mission of the institution. This in turn will shape the strategies that a particular department will adopt for its curriculum and teaching.

More than any other researcher in a department of chemistry, the scholarly work of the chemical education researcher will be influenced by both the research and educational missions of the department and of the institution. The research mission of the department will have an impact on the expectations that will exist for scholarly work in the area of chemical education. It is not fair to oversimplify the situation by saying that institutions with less research emphasis will expect less research from a chemical education researcher and those with a heavy research emphasis will expect more. Indeed, many departments with a heavy research emphasis do not consider their chemical education researchers to be researchers in chemistry, and in some cases do not expect them to be. For example, there are institutions who hire a specialist in chemical education to direct a large undergraduate education program, sometimes consisting of many hundreds of students who will be enrolled in first and second year chemistry courses with laboratory components. Likewise, smaller institutions, where there is less traditional chemistry research, may be interested in emphasizing research in chemical education due to its relatively smaller need for laboratory space and overhead costs.

There are nearly 30 institutions nationwide offering Masters and Doctoral degrees in Chemical Education today. This level of commitment to the graduate program in chemical education indicates a clear expectation for research productivity, at some level, on the part of the faculty in those departments. This mode, in fact, is becoming increasingly common at higher education institutions, though by no means ubiquitous. Therefore, the impact that the research mission of a particular department will have on the scholarly work of a chemical education researcher varies greatly across institutions, depending on the role the department expects for that person, as will be discussed in greater detail below.

The teaching mission will also play a role in shaping the scholarly work of the researcher. There are two ways in which this can potentially occur, namely, providing a "laboratory" for the educational research and serving as the crucible whereby the scholarly activities of the reearcher achieve credibility within the department. Not every chemical education researcher will specialize in research activities that focus on the population of students that are taught within their own departments. For example, researchers may specialize in K-12 educational research, in-service teacher educational research or other subjects that lie outside the classroom setting of the researcher's own institution. However, a large fraction of researchers in chemical education today do focus, to some degree, on research questions that are directly related to the population of students served by their own departments and institutions. This makes sense, given that the sample population of interest and associated data are easily accessible, reducing the cost and complexity for carrying out the research. When this is the case, then the type of research questions will be highly influenced, of course, by the population at hand and the type of teaching program that the department supports.

Perhaps less obvious and more intriguing is the second method by which the teaching mission can influence the scholarly work of the chemical education researcher. Even in a department where the primary emphasis is on research, the area of specialty of a chemical education researcher is education. This means that the classroom can be an "arena of opportunity" in relationship building with other chemists in the department. The experience of the chemical education researcher can result in teaching methodologies, assessment practices, and even curricular structuring that are novel for the department. When chemical education researchers utilize approaches that they understand as a result of their own research, this can serve to underscore the validity and usefulness of their research area or their own work. In other cases, the researcher can serve as a steward to introduce these approaches to their colleagues. In either scenario, the chemical education researcher does face a level of scrutiny in their own teaching to which other faculty in the department may not be subjected. In essence, the department may see this person's teaching as a lens for understanding the field of chemical education research, often a new undertaking for departments of chemistry. This is a large responsibility for chemical education researchers to bear in their own teaching, but it is part of the foundation for the communication that is crucial in this field and which was alluded to in the opening chapter of this volume.

Goals of the Chemical Education Researcher

There are numerous ways in which the scholarship of education can be defined. In 1990, Ernest L. Boyer's report *Scholarship Reconsidered: Priorities of the Professoriate* by the Carnegie Foundation for the Advancement of Teaching (*1*) provided an expanded view of what was defined as scholarship. This report was further interpreted with respect to chemistry education research in a document by the Task Force on Chemical Education Research of the ACS Division of Chemical Education (*2*). In particular, Boyer's report provided for scholarship to be defined along four dimensions: discovery, teaching, integration and application (Table I).

Table I. Dimensions of Scholarship

Dimension	*Description*
Discovery	Creative, investigative scholarship seeking new understanding.
Teaching	Creative, investigative and proactive engagement in education through the application and exploration of various pedagogical approaches.
Integration	Synthesis of knowledge by developing meaning across disciplines and placing findings within a larger context.
Application	Using knowledge to inform, address or solve broader issues. Engagement in these broader issues.

SOURCE: Reproduced with permission from reference 1. Copyright 2000 John Wiley & Sons.

Traditional chemistry research is thought of as existing within the discovery dimension. Research in chemical education can also exist within this dimension. The overall approach to research is the same for chemical education researchers and other scientists. Chemical education researchers study hypotheses using a research design with instruments to collect data and apply theory to analyze their data and interpret the results. The exact techniques and tools are different (*3-6*), but the scientific approach is intact, such that the scholarship of discovery is an appropriate description for this type of work.

However, there are other modes in which the work of a chemical education researcher can be considered scholarship yet not fall within the discovery dimension. In fact, this is also true for researchers in other areas of chemistry. The American Chemical Society has revised its own statement on scholarship (7) to state:

> *In addition to discovery research, scholarship in the chemical sciences and engineering includes the integration, application, and teaching of chemical sciences and engineering principles and practices.*

The scholarship of teaching encompasses, broadly, much of what we consider to be chemical education research: transforming and extending knowledge creatively (*1*). This type of scholarship also includes work that contributes to an understanding of established and development of new approaches to effective teaching.

The last two dimensions listed by Boyer also describe the nature of chemical education research. The scholarship of integration considers ways in which different disciplines inform and contribute to each other. There are numerous approaches to chemical education research that draw from theories and methods common to anthropology, psychology, cognitive sciences, computer sciences, and statistics. While these may not proceed by the experimental paths that chemists are accustomed to seeing, they are nonetheless valid and useful scientific approaches for answering certain research questions.

Finally, the scholarship of application encompasses those areas of chemical education that focus on development of materials, curriculum, and instructional methodologies, as well as implementation in the classroom. This is a highly applied facet of chemical education research. An analogy can be made with research in other subdisciplines of chemistry, in that some areas are more fundamental in nature, and others are much more applied, yet the entire spectrum is necessary for the chemical research enterprise. A different way to interpret the scholarship of application is in the arena of outreach, whereby chemistry knowledge generated through the traditional research subdisciplines is translated and communicated to the larger community.

Ultimately, chemical education researchers should determine not only their specific research questions, but also the type of scholarship that their work represents. The dimensions of scholarship outlined above can be used to assist in making this determination. Doing so provides a clear way to communicate the goals of the research agenda to administrators and other scientists in the department. It also simplifies the task of mapping the goals of the research onto those of the department to ensure that there is a reasonable fit, with respect to both the research agenda and the role that the researcher can play in the department as a whole.

Fostering Interactions where Goals Overlap

Academic research in science is changing in response to forces such as global competitiveness, emerging technological and scientific interests – sometimes in response to political, environmental or health crises – and availability of funding. A research approach that is interdisciplinary and collaborative is becoming more common and more feasible than at any time in the past, given the ease of travel, information exchange, and distance interaction that now exists. Similarly, these forces are having an impact on the way that universities carry out their work. The research and educational missions of universities, and even of funding agencies, are beginning to require increased overlap and articulation between the two types of activities. Take, for example, the statement of "Broader Impacts" that now accompanies all National Science Foundation grant proposals and their reviews (8). This statement requires that all proposals to the agency address issues such as "how well does the activity advance discovery and understanding while promoting teaching, training and learning?" Additional examples of this type of broadened scope for proposals to funding agencies were described by Sawrey in chapter 13 of this volume (9).

In addition to funding, there are various mandates that act upon universities and departments through federal, regional, or disciplinary accreditation. These also are looking for an enhanced connection between the research and teaching missions of institutions and departments. The ACS is proposing new guidelines for approval of undergraduate chemistry programs. The proposal by the Committee on Professional Training for the new guidelines (10) states that "opportunities for original undergraduate research...are highly recommended." The proposed guidelines emphasize that undergraduate research can be used to qualify as in-depth coursework for ACS approval of the curriculum, and that "because of its importance for the education of chemistry majors, the opportunity for undergraduate research should be offered whenever possible." An additional example comes from the criteria for higher education accreditation of the North Central Association (11), which were revised in 2003. In the section which most directly discusses research – the criterion on "acquisition, discovery and application of knowledge" – a core component states that "acquisition of a breadth of knowledge and skills and the exercise of intellectual inquiry are integral to its educational programs."

Researchers in chemical education working closely with other faculty and with administrators in chemistry departments can be instrumental in creating bridges between the research and educational goals of departments. A characteristic of many researchers in chemical education is that they have a solid grounding in the science discipline itself. Yet they have additional expertise in the issues, techniques, and findings of educational research. Their own work, regardless of the specific question or topic, will naturally overlap in research and education. But there are additional ways in which the endeavors could have direct benefit to other researchers in the department and the educational researcher. Sawrey (9) has described some ways to do this. There are excellent opportunities for chemical education researchers to be associated with the broader impacts components of grant proposals. There are also strategies that can be used to infuse findings from chemical research into the curriculum, such as materials development projects. Projects in chemical education can also be used to carry the findings of chemical research to populations outside the undergraduate chemistry majors or outside the department, such as through adapt-and-implement or informal education projects.

Undoubtedly, there are educators in each chemistry department who take a serious interest in teaching even though their own research interests are in the traditional chemical subdisciplines. It is likely that many of these people are eager to incorporate approaches and materials into their classes that will provide a high quality experience for their students. Collaborations with faculty colleagues who share these interests are an excellent opportunity for a chemical education researcher to strengthen ties throughout the department as a whole, through the faculty network. Education researchers are in a wonderful position to build these collaborations, not only within a single department, but also across scientific disciplines and academic units (departments/divisions/colleges). Education is a common thread to almost all the academic departments at higher education institutions, and research endeavors that meet common interests can be quite fruitful.

With some creativity on the part of the researchers and a willingness of colleagues to collaborate, there are numerous approaches that can be taken that will allow for a chemical education research program to become an integral part of a chemistry department's mission. It is wise to be informed about the various requirements that chemical researchers and chemistry departments need to meet so that opportunities for overlap can be identified early. It is also important that chemical education researchers examine their own goals with respect to their roles both in their departments and in any collaborations. It is possible for researchers in chemical education to find themselves in the role of "the assessment person" or the person in charge of curriculum decisions without necessarily intending to take on that role. These roles are perfectly valid and useful for a faculty member if it is what he/she seeks for his/her own professional direction. However, if it is not, then it is very important that the chemical education researcher define an inquiry of interest to him/herself professionally when developing a collaboration of this nature. If tenure and peer-reviewed publications are part of this person's professional goals, then the chemical education researcher will need to be able to demonstrate what his/her scholarly contributions were to various projects. In the end, the department as a whole and the researcher can best be served when there is professional growth for the chemical education researcher in his/her departmental or other collaborations.

Broadening the Horizons of Other Stakeholders

We come full circle in this volume to the issue of communication between chemical education researchers and other chemists, as was the message of the opening chapter. Research in the field of chemical education, especially within departments of chemistry, is a relatively new endeavor. The oldest degree-granting programs in the nation in this area are no more than 25 years old. As a result, there are many people who are not aware of the work that is carried out in this field, the methods its practitioners employ, nor its potential as a research area in a chemistry department. Indeed, that is one impetus for the development of this volume in the first place. In addition to a general unfamiliarity with the field, the reality is that there are many scientists and administrators who may cast a skeptical eye on chemical education as a research field. When there are limited resources to be shared, and long-established, possibly narrow, beliefs about what constitutes research, it may take time and continued championing to help others modify their view of the value and validity of chemical education research. Professionalism, collegiality and a sharing of information are key to this effort.

It is important to realize that there are numerous movements occurring nationally and internationally that are putting a greater focus on education at higher education institutions. The report on the future of higher education by the commission appointed by Secretary of Education, Margaret Spellings (*12*), may be an indicator of increasing needs to communicate the details of different types of educational approaches to both the professoriate and to the public. The report outlines various failures in K-16 education in the United States, and indicates that the lack of institutional accountability in higher education is compounding all of these problems. It calls for mechanisms by which "reliable information about the cost and quality of postsecondary institutions" will become available. Institutions that have traditionally focused very heavily on their research mission, especially due to its important role in the financial viability of many of its programs, may find themselves facing tremendous new challenges as a result of requests of the type that the Spellings report makes. The direct consequences of federally mandated efforts, such as the Spellings Commission report, will likely affect only public institutions initially. However, if this results in major programmatic changes at public institutions, even private ones may find it necessary to provide similar transparency for those who are interested in their programs. Chemical education researchers, and educational researchers in other science and technical disciplines, are in an excellent position to be primary sources for the type of information that is needed. By carrying out research on student learning, teaching methodologies, and the impact of instructional innovations, these researchers can report in a truly scientific manner on educational efforts. As a result, researchers in chemical education and other science education fields can be instrumental in assisting with the departmental response to required changes in process or curriculum.

One key to effectively communicate work in this field is to consider the needs of the audience, which is both wide and varied. Chemical education researchers must report research findings in peer-reviewed journals to experts in educational research. Precise language and thoroughly developed methodologies are cornerstones of such communication. However, there is also an audience of non-expert *consumers* of the research carried out in this field. The audience that can be directly and immediately impacted by findings in chemical education research is arguably larger, and more varied, than for any other single subdiscipline of chemistry. This audience spans a range that includes policymakers, educational administrators, educators, scientists, parents, and students. The breadth of this audience is both a challenge and an opportunity for the field of chemical education. Hence, there is a great need to find multiple avenues and modes for communicating the findings of educational research. It is imperative that chemical education researchers take the lead in representing our work to others. Finding the correct balance between lay language and accurate representation of the work and findings in the field can be a challenging task, but it is one that can best be met by those who are experts in chemical education research and understand its intricacies as well as its broader implications.

References

1. Boyer, E. L.; *Scholarship Reconsidered: Priorities of the Professoriate*; John Wiley and Sons: New York, NY, 1990.
2. Bunce, D.; Gabel, D.; Herron, J. D.; Jones, L. *Journal of Chemical Education*, **1994**, *71*, 850.
3. Bretz, S. L.; "Qualitative Research Designs in Chemistry Education Research." In *Nuts and Bolts of Chemical Education Research*; Bunce, D. M., Cole, R., Eds.; American Chemical Society: Washington, DC, 2007.
4. Sanger, M. J.; "Using Inferential Statistics to Answer Quantitative Chemical Education Research Questions." In *Nuts and Bolts of Chemical Education Research*; Bunce, D. M., Cole, R., Eds.; American Chemical Society: Washington, DC, 2007.
5. Towns, M. H.; "Mixed Methods Approaches in Chemical Education Research." In *Nuts and Bolts of Chemical Education Research*; Bunce, D. M., Cole, R., Eds.; American Chemical Society: Washington, DC, 2007.
6. Scantlebury, K.; Boone, W. J.; "21st Century Statistics for 21st Century Studies." In *Nuts and Bolts of Chemical Education Research*; Bunce, D. M., Cole, R., Eds.; American Chemical Society: Washington, DC, 2007.
7. Education and International Activities Division, American Chemical Society, Chemunity News, Fall 2001, pg. 17.

http://www.chemistry.org/portal/resources/ACS/ACSContent/education/chemunitynews/chemunity_news_fall01.pdf (Accessed 10-31-06)
8. National Science Foundation, *Broader Impacts Showcase.* http://www.nsf.gov/pubs/2005/nsf0540/nsf0540.jsp. (Accessed 9-26-06).
9. Sawrey, B. A.; "Collaborative Projects: Being the Chemical Education Resource." In *Nuts and Bolts of Chemical Education Research*; Bunce, D. M., Cole, R., Eds.; American Chemical Society: Washington, DC, 2007.
10. Committee on Professional Training of the American Chemical Society. *Proposed Revision of the ACS Guidelines for Undergraduate Chemistry Programs.* http://www.chemistry.org/portal/resources/ACS/ACSContent/education/cpt/ACS%20Proposed%20Guidelines%20Revision. (Accessed 10-2-06).
11. Institutional Accreditation: An Overview; Higher Learning Commission of the North Central Association of Colleges and Schools: Chicago, IL, 2003. http://www.ncahlc.org/download/2003Overview.pdf. (Accessed 10-2-06)
12. A Test of Leadership: Charting the Future of U.S. Higher Education; Commission on the Future of Higher Education, appointed by Secretary of Education Margaret Spellings; U.S. Department of Education: Washington D. C., 2006. http://www.ed.gov/about/bdscomm/list/hiedfuture/reports/pre-pub-report.pdf. (Accessed 9-26-06).

Indexes

Author Index

Abraham, Michael R., 47
Bauer, Christopher F., 183
Boone, William J., 149
Bretz, Stacey Lowery, 79
Bunce, Diane M., 1, 35
Cole, Renée S., 1, 183
Cooper, Melanie M., 171
Jones, Loretta L., 19
Sanger, Michael J., 101

Sawrey, Barbara A., 203
Scantlebury, Kathryn, 149
Scharberg, Maureen A., 19
Towns, Marcy Hamby, 135
VandenPlas, Jessica R., 19
Walter, Mark F., 183
Weaver, Gabriela C., 215
Williamson, Vickie M., 67
Zare, Richard N., 11

Subject Index

A

Adapting existing survey and test instruments, 160–162
Advanced organizers, 56
Analysis of covariance tests (ANCOVA), 116–120
Analysis of variance tests (ANOVA), 115–120
Anchoring in Rasch analysis, 165–166
ANCOVA. *See* Analysis of covariance tests
Anecdotal evidence in chemical education research, 175
Animations, benefits in student understanding, chemistry topics, 73–74, 75–76
ANOVA. *See* Analysis of variance tests
Assessment goals, 185
Assessment of student learning, 183–201
Attitudes, learning goals, assessment, 190t–191t
Audit trails, data quality in qualitative methodologies, 95
Ausubel's subsumption theory, 56–58

B

Between-treatments degrees of freedom in F-statistics calculations, 115
Bias minimization in chemical education research, 109–111
Budget development for large projects, 211–212
Budget establishment for proposals, 28–29

C

Camille and Henry Dreyfus Foundation, proposal requirements, 27–28
Case study methodology, qualitative research tradition, 89
Categories, data analysis. *See* Coding and categories
Causality, questionable conclusions in research studies, 173
CER. *See* Chemical education research
Checklists for writing research questions, 45
Chemical education
　sources of theory, 54–55
　theories and research, 56–63
Chemical education and chemistry communities, relationship, 215–225
Chemical education literature, mixed methods research designs, examples, 145–146
Chemical education research
　funding, 19–33
　inferential statistics, 101–103
　introduction, 1–10t
　mixed methods design, 135–148
　qualitative research designs, 79–99
　test and survey design, 149–169
Chemical education researcher, goals, 219–221
Chemical universe divisions, 12
Chemistry as discipline, contribution to chemical education theory-base, 54

Chemistry departments
  characteristics affecting chemical education research, 217–219
  practical considerations in qualitative research, 95–96
  relations between chemistry and chemical education communities, 215–225
Chemistry education research. *See* Chemical education research
Chi-square goodness-of-fit tests, 121–122
Chi-square tests for nominal data, 121–126
Chi-square tests of homogeneity, 122
Chi-square tests, uses and assumptions, 123–126
Choice of statistical tests, guidelines, 126–128t
Choices in mixed method design, decision tree, 138f–139f
Clearinghouses for grant information, websites, 24–25
Coding and categories, data analysis in qualitative methodologies, 92–93
Coincidental correlations, 113
Collaboration incorporation for proposal preparation, 29
Collaborations with
  Community partnerships, 208
  Inter-institutional projects, 208–209
  K-12 sector, 207–208
  Schools of education, 206–207
  Science and engineering faculty, 204–206
Collaborative projects, 203–214
  benefits and cautions, 212–213
Communication
  between chemical education researchers and other chemists, 223–224
  decision-making, and laboratory skills, learning goals, assessment, 195t

Community partnerships with chemical education researcher, 208
Concept mapping, 56, 57f
Conceptual change theory, 61
Conclusions from education experiments, 171–182
Concurrent nested strategy in mixed research design, 143–144f, 146
Concurrent triangulation strategy in mixed research design, 142–143f
Consistency in surveys and test items, 157–159
Construct development in survey and test design, 153–154
Contingency tables, 122
Correlation coefficients, 112–113
Covariance analysis. *See* Analysis of covariance
Credibility and member checks, data quality in qualitative methodologies, 94
Cross-age studies, student misconceptions on particulate nature of matter, 70, 71

D

Data analysis, 195–197
  and integration in mixed methods research, 144–145
  in qualitative methodologies, 91–93
Data collection, 193–195
  design approach, 188, 192
  in qualitative research designs, 82–88
Data quality in qualitative methodologies, 93–95
Data selection determination, 192–193
Degrees of freedom
  reporting in F-statistics calculations, 115–116
  student t-tests, sample size effects, 114

Department of Education (US) as funding source, 23–24
Dependability, confirmability, and audit trails in qualitative methodologies, 95
Dependent measures ANOVAs. *See* Repeated measures analysis of variance tests
Dependent measures t-tests. *See* Repeated measures t-tests
Dependent tests of proportions. *See* Tests of proportions
Descriptive statistics (definition), 106
Directional research hypotheses, concerns, 104
Directorate of Undergraduate Education (DUE), 22, 23
Disequilibration in Piaget's theory of intellectual development, 59
Document analysis, data collection in qualitative research, 86
DUE. *See* Directorate of Undergraduate Education

E

Education schools, working with chemical education researcher, 206–207
Educational theories for chemical education, 56–63
Effect size (definition), 107
Error probabilities, inferential hypothesis testing, 106–107
Ethnography, qualitative research tradition, 90–91
Exemplary papers in chemical education research, 177–180
Experimental method, relationship to question, 44
Exploratory data analysis. *See* Post-hoc research questions

External validity, 109

F

Feasibility, researchable questions, 39–40
Federal agencies as funding sources, 19–24
Fieldnotes for data collection in qualitative research, 87–88
Final questions, 44–45
Finalizing the proposal, 29–30
FIPSE. *See* Fund for Improvement of Post-secondary Education
Focus on questions in educational research, 50–51
Formulation of questions, 42–45
Functioning model, Piaget's theory of intellectual development, 58–61
Fund for Improvement of Post-secondary Education (FIPSE), 24
Funding, chemical education research, 19–33
Funding source location, 19–25

G

Gagné, 58
Gap between learning research and educational practice, reasons for, 52–53
Good questions, components, 36–42
Grant proposals for chemical education research, 6
Grants for chemical education research, classifications, 19–20
Grounded theory, qualitative research tradition, 91
Guidelines for choice of statistical tests, 126–128t
Guidelines for this book's use, 7–10t

H

Hawthorn effect, 174
Human subjects in chemical education research, 4–5
    See also Institutional Review Board
Hypothesis testing results, 105–106

I

Implementation step in mixed method designs, 137
Independent assessment instrument design, 197–198
Independent measures t-tests, 114
Independent test of proportions. See Tests of proportions
Inferential hypothesis testing, error probabilities, 106–107
Inferential statistics (definition), 106
Inferential statistics in quantitative chemical education research, 101–133
Information gained from student assessments, use, 185–186
Informed consent, 84
Inquiry-based instructional strategies, 59
Institutional Review Boards (IRB), 5, 83–84, 85, 107, 174, 197, 210–211
    See also Human subjects research
Integration of approaches in mixed method designs, 140
Intellectual development, Piaget's theory, 58–61
Inter-institutional projects, 208–209
Interaction plots, 117–119
Interactions when goals overlap, 221–222
Interactivity, learning goals, assessment, 193t
Internal validity, 108–109
Interviews, data collection in qualitative research, 84–85
IRB. See Institutional Review Boards
Item fit and person fit, Rasch statistics, 163, 165
Item length in surveys and tests, 154
Item maps, Rasch analysis software, 163–164f
Item style and word choices in surveys and tests, 152–153

J

Just-in-Time Teaching (JITT), 56

K

K-12 sector, collaboration with chemical education researcher, 207–208
Knowledge learning goals, assessment, 187t

L

Learning assessment, 5
Learning cycle approach, relation to Piaget's functioning model, 59–60t
Learning goals for knowledge, 187t
Learning preferences and learning theory, 51–52
Length constraints for surveys and tests, 159
Letters of intent, 26
Letters of support for collaborative proposals, 211–212
Level of significance (definition), 106–107
Literature, research results, interpretation, 131
Literature reviews in theory-based research studies, 104
Locating funding sources, 19–25

M

McNemar's tests for significance of change, 122
Measuring variables, researchable questions, 40–41
Member checks, data quality. *See* Credibility and member checks
Mental models, particulate nature of matter, 68–69
"Merit review criteria" for proposals submitted to National Science Foundation, 27
Metacognition learning goals, assessment, 189t
Methodologies, choosing qualitative vs. quantitative, 80–82
Misconception/alternative conception studies, particulate nature of matter, 69–72
"Misfit." *See* Item fit and person fit
Mixed method designs, key decisions, 136–140
Mixed method study, exemplary paper in chemical education research, 179–180
Mixed methods in chemical education research, 135–148
data analysis and integration, 144–145
definition, 136
Multiway frequency analysis, 122–123

N

Narrative writing for proposals, 26–28
National Institutes of Health (NIH) as funding source, 24
National Science Foundation (NSF) as funding source, 22–23
"merit review criteria," 27
NIH. *See* National Institutes of Health
Non-parametric tests. *See* Chi-square tests
Non-validated instruments in chemical education research, 175
Novak, Joseph, concept mapping, 56, 57f
NSF. *See* National Science Foundation
Null hypothesis evaluation, 126–127, 129
Null hypothesis statement, 105–106

O

Observation methodologies, data collection in qualitative research, 85–86t
OESE. *See* Office of Elementary and Secondary Education
Office of Elementary and Secondary Education (OESE), 23
Office of Innovation and Improvement (OII), 23
Office of Postsecondary Education (OPE), 23, 24
Office of Science Education (OSE), 24
OII. *See* Office of Innovation and Improvement
One-sample t-tests, 114
One sample test of proportions. *See* Tests of proportions
OPE. *See* Office of Postsecondary Education
OSE. *See* Office of Science Education
Overgeneralization in chemical education research, 176–177
Overlapping goals, chemists and chemical education researchers, 221–223

P

"Paper and pencil" test development. *See* Quality items in surveys and tests
Particulate nature of matter, impact of theory-based research on student understanding, 67–78
Pearson product-moment correlation coefficient, 11–1132
Permission for student data collection. *See* Institutional Review Board
Person fit and item fit, Rasch statistics, 163, 165
Phenomenology, qualitative research tradition, 89–90
Philosophy, contribution to chemical education theory-base, 55
Physical layout, surveys and tests, 159
Piaget's theory, intellectual development, 58–61
Piloting of survey and test instruments, 159–160
Planning effective student assessment, considerations, 184–197
POGIL. *See* Process-oriented guided-inquiry learning
Pooling items in surveys and tests, 150–152f
Population selection in research studies, 173–174
Post-hoc comparisons, 117
Post-hoc research questions, 131
Power, statistical comparison (definition), 107
Practical significance, 129–130
Preparation for writing the proposal, 25–26
Priority of approaches in mixed method designs, 137, 140
Private agencies as funding sources, 20, 21t
Problem identification, researchable questions, 38–39
Process-oriented guided-inquiry learning (POGIL), relation to Piaget's functioning model, 60
Proposal finalization, 29–30
Proposal writing, 25–29
Psychology, contribution to chemical education theory-base, 54–55
Pygmalion effect, 174

Q

Qualitative research designs, chemistry education research, 79–99
data collection, 82–88
Qualitative research in chemistry department, practical considerations, 95–96
Qualitative research tradition selection, 88–91
Qualitative study, exemplary paper in chemical education research, 178–179
Quality items in surveys and tests, 152–162
Quantitative chemical education research, inferential statistics, 101–133
Quantitative study, exemplary paper in chemical education research, 177–178
Questions for research, defining and constructing, 35–46
Questions from chemistry community on teaching chemistry, 11–18
Questions in chemical education research, 2–3
Questions in writing mechanics, 42–45

R

Rasch analysis software, 167–168
Rasch model for test and survey development, 162–168
Rating category development, 155–156
Rating scale evaluation in Rasch analysis of data, 166
Raw score-equal interval conversion table in Rasch analysis software, 166
Recurring themes in chemical education research, 2–5
REESE Program. *See* Research and Evaluation on Education in Science and Engineering
Relationship building, chemists and chemical education researchers (overview), 216–217
Reliability, 107–108
Repeated analysis of variance tests, 115–116
Repeated (dependent) measures t-tests, 114–115
Repeated (dependent) tests of proportions, 120
Replication studies, 130
Reporting results, 129–131
Representative sampling, 108–109
Request for proposal (RFP), importance of following, 26–27
Research and Evaluation on Education in Science and Engineering (REESE) Program, 22–23
Research design, planning and implementation, 107–111
Research hypothesis development, comparison to research questions, 102–104
Research methodologies in chemical education research, 3–4
Reversed items in surveys and test development, 156–159
RFP. *See* Request for proposal

S

SALG. *See* Student Assessment of Learning Gains
Sample size effects, degrees of freedom, student t-tests, 114
Sampling strategy for thick description production, 82–83
Scaffolding, 62
Scholarship, American Chemical Society statement, 220
Scholarship dimensions, 219–220
Science and engineering faculty, working with chemical education researcher, 204–206
Scientific investigation, basis, 45
Selection of qualitative research tradition, 88–91
Self-reported learning in chemical education research, 175–176
Self selection in research studies, 173–174
Semi-structured interview guide, data collection in qualitative research, 84
Sequence problem in curricular material presentation, 61–63
Sequential explanatory design in mixed research design, 141–142f, 145
Sequential exploratory design in mixed research design, 140–141, 146
Shaw, George Bernard, "Maxims for Revolutionists," 13, 16
Significance level definition, 106–107
Social constructivism, Vygotsky's, 61
Sociology, contribution to chemical education theory-base, 55
Software programs for Rasch analysis, 167–168
Spellings (Margaret) Commission report, 223
Stage model, Piaget's theory of intellectual development, 58
Standardized effect size, 107

Standardized residuals, 121–122
Statistical significance, 129–130
Statistical tests, guidelines for choice, 126–128t
Student Assessment of Learning Gains (SALG), 176
Student learning assessment, 183–201
Student t-tests, 113–115
"Students learn differently" position, 51–52
Subjects and verbs in writing researchable questions, 43
Subjects for study, researchable questions, 40
Subsumption theory (Ausubel's), 56–58
Survey and test design for chemistry education research, 149–169

T

Take home messages, researchable questions, 41–42
Task Force on Chemistry Education Research, ACS Division of Chemical Education, statement on student learning, 94
Test and survey design for chemistry education research, 149–169
Test "distractors" construction, 154
Test statistic choices, 111–128t
Tests of proportions, 120–121
Theoretical educational perspective in collaborative projects, 209–210
Theoretical frameworks
  data analysis in qualitative methodologies, 92
  identification and application, 187–188
  importance for research, 47–66
  in mixed method designs, 140
Theories and research in chemical education, 56–63

Theory-base development, roadblocks, 52–53
Theory-based questions, importance, 42
Theory-based research, influence on teaching practices, 74–76
Theory vs. empiricism, controversy, 49–50
Thick description and transferability, data quality in qualitative methodologies, 93–94
Thick description produced by purposeful sampling strategy, 82–83
Timeline for writing the proposal, 26
Transferability and thick description, data quality in qualitative methodologies, 93–94
Treatment/intervention studies, student understanding, particulate theory of matter, 72–74
Triangulation, research methods in data collection in qualitative research, 86–87
Twenty questions for chemical education researches, 14–15
Two-way analysis of variance tests, 116
Type I error (definition), 106–107
Type II error (definition), 106–107

U

Unarticulated learning theory, 52
Undergraduate chemistry programs, American Chemical Society guidelines, 221

V

Validity, 107
Variables, intervening or confounding, in questions, 44

Variance analysis. *See* Analysis of variance
Verbs and subjects in writing researchable questions, 43
Vygotsky's social constructivism, 61

W

Winsteps software program for Rasch analysis, 167–168

Within-treatments degrees of freedom in F-statistics calculations, 115
Word choice in writing researchable questions, 43
Writing mechanics, researchable questions, 42–45
Writing the proposal, 25–29

Z

Zone of Proximal Development, 61